D1190499

Learning to Communicate in Science and Engineering

Learning to Communicate in Science and Engineering

Case Studies from MIT

Mya Poe, Neal Lerner, and Jennifer Craig

foreword by James Paradis

The MIT Press
Cambridge, Massachusetts
London, England

For information about special quantity discounts, please e-mail special_sales@mitpress.mit.edu

This book was set in Stone Serif and Stone Sans on 3B2 by Asco Typesetters, Hong Kong.
Printed and bound in the United States of America.

Library of Congress Cataloging-in-Publication Data

Poe, Mya, 1970–
Learning to communicate in science and engineering : case studies from MIT / Mya Poe, Neal Lerner, and Jennifer Craig ; foreword by James Paradis.
p. cm.
Includes bibliographical references and index.
ISBN 978-0-262-16247-0 (hardcover : alk. paper)
1. Communication in science—Case studies. 2. Communication in engineering—Case studies. 3. Writing, Humanistic. I. Lerner, Neal. II. Craig, Jennifer, 1945– III. Title.
Q223.P64 2010
501′.4—dc22 2009024788

10 9 8 7 6 5 4 3 2 1

Contents

Foreword

Although writing and speaking skills are stock-in-trade for the practicing scientist or engineer, communication is by no means an easy subject to teach in the university. It helps, no doubt, to have students write and speak as much as possible across the university curriculum. Yet communicating science and engineering, the authors of this detailed and fascinating study show, is also a cognitive and social process that is discipline specific. It flows from a professional identity and, when effective, gives persuasive force to scientific and technological findings while placing them into the framework of a recognizable field.

One challenging but creative approach to teaching students to communicate in their disciplines is to see if we can make learning to communicate an integral part of learning the disciplinary subject. The case studies in this book take the teaching of communication into a series of undergraduate subjects in these majors at MIT: biology, biological engineering, aeronautical and astronautical engineering, and biomedical engineering. Analyzing the writing and speaking development of seventeen students in seven MIT science and engineering classes (in five departments), Poe, Lerner, and Craig explore ways in which learning to communicate helps students identify with and think in a discipline. Effective communication, they show, is closely linked with an emerging disciplinary identity, and certain communication tasks with feedback will strengthen this important connection: practice with forms (genres), working with peers in team projects, and receiving feedback from subject instructors and peers. These case studies, then, can be seen as both detailed explorations of how students learn to meet the expectations of communicating in specific science and engineering disciplines and as a series of best practices that provide educators with clear strategies for improving teaching communication in the disciplines.

The collaborative model of teaching communication showcased in this study has a long history at MIT. Although evaluation of technical reports was a practice of science and engineering education at MIT as early as the 1890s (Russell 2002), Robert R.

Rathbone's efforts in the early 1950s began MIT's modern era of integrating communication instruction into various core science and engineering laboratory subjects. Rathbone, who had been hired as a report writer for Project Whirlwind by Jay Forrester, the MIT inventor of the computer magnetic core memory, developed a popular report writing course in the early 1960s for undergraduate science and engineering majors. He also had a thriving consulting practice as a short-course instructor in industry teaching report writing to scientists and engineers. Extending his collaborations with Forrester to other MIT engineering faculty, Rathbone began experimenting with bringing this real-world short-course experience back into the engineering classroom in a variety of presentations on communication, help with syllabus design in the core engineering subject, and feedback on student written and oral presentations.

The establishment in the mid-1970s of the independent MIT Writing Program in the School of Humanities and Social Sciences gave new force to these collaborations, which Rathbone and his colleagues expanded to a variety of core laboratory subjects throughout the departments of the School of Engineering. This effort, which was increasingly funded by the School of Engineering, demonstrated not only that engineering faculty valued good communication and were interested in experimenting with new models of communication education, but also that a consulting or distributed model of communication instruction could be sustained through a network of collaborative interdepartmental structures.

Over the following twenty-five years, after extensive experimentation and development by various members of the MIT Writing Program, much of it spearheaded by Les Perelman and the coauthors of this well-researched, probing study, the Institute Communication Requirement was developed and passed by an MIT faculty vote in 2000. Perhaps the most compelling of the many features of this requirement was the mandate that communication instruction in the respective major be taken by every MIT undergraduate in both the junior and senior years. The Writing Across the Curriculum Program, a division of the Program in Writing and Humanistic Studies, became the primary source of this collaborative instruction. Poe, Lerner, and Craig take us deep into the territory of the collaborative teaching enterprise, where communication instructors negotiate with subject instructors to design curricular interventions in communication that seek to improve student learning of the subject matter. What I find absorbing about these accounts is their essential honesty about the messiness of the undertaking, with all its conflicts, misunderstandings, lack of closure, and variability. There are questions of authority between subject and communication instructors that do not always get resolved. The reassuring control of one's own classroom is relinquished, in exchange for a kind of opportunistic pedagogy seeking to insert itself at crucial junctures in the harried process of a demanding learning curve. Follow-through is a constant problem. Timing can be a nightmare. Students sometimes resist working with the communication instructor, the subject instructor, or each other. The measured world of the communication textbook or even the luxury of one's own classroom space seem distant. And yet these situated learning spaces provide what is undoubtedly a dynamic locus for teaching students to communicate in the disciplines. They come close to the near chaos of real world multitasking, where learning, thinking, and communication tasks are often completed with colleagues just in time.

This book takes a major step forward in both the scholarly study and the pedagogy of writing in the science and engineering disciplines. Its chapters are framed in the most up-to-date research in writing studies, but the research is thoughtfully subordinated to the object of improving our understanding of the classroom situation. It offers a new window on the complex relationship between communication and the practice of science or engineering, and it provides many concrete strategies for improving the teaching of communication in science and engineering fields where multiple authorship is the norm. As a writing program administrator who has had the privilege of seeing Rathbone's wise and entertaining one-man communication show mature into the substantial, far-ranging collaborative pedagogy of Poe, Lerner, and Craig and their many MIT colleagues in science, engineering, and communication, I highly recommend this book to science, engineering, and communication educators everywhere.

James Paradis
Head, Program in Writing and Humanistic Studies
Professor of Science and Engineering Communication
Robert M. Metcalfe Professor of Writing
Massachusetts Institute of Technology

Acknowledgments

This book would not have been possible without the support of our colleagues at MIT. Special thanks to Leslie Perelman for his long-standing stewardship of the Writing Across the Curriculum Program and his collegial guidance. Thanks also to James Paradis, the head of the Program in Writing and Humanistic Studies, for his encouragement and support of this project.

To the faculty and instructor colleagues in the disciplines who graciously allowed us to study their practices and describe them in this book, we owe a tremendous debt. In the Department of Biology, we thank Marilee Ogren, Dennis Kim, Thomas Schwartz, and Stephanie Capaldi. In Biological Engineering, we thank Natalie Kuldell and Drew Endy. In Health Sciences Technology, we thank Sangeeta Bhatia and Martha Gray. In the Department of Electrical Engineering and Computer Science, we thank Dennis Freeman, Jongyoon Han, Tom Weiss, and Joel Voldman. In the Department of Aeronautics and Astronautics, we thank Edward Greitzer, Sheila Widnall, Ian Waitz, David Darmofal, Earll Murman, David Miller, John Keesee, Jane Connor, John Hansman, Brad Holschuh, Dick Perdichizzi, David Robertson, and Todd Billings in addition to the exceptional support and administrative staff members that keep us all going. In addition to instructional staff, we are indebted to the student participants in this study, and envision long, fulfilling careers for all, whatever the types of communication practices they are engaged in.

MIT Press has been generous with its support for this project. Thank you to our editors, Doug Sery, Katie Helke, and Sandra Minkkinen. Thank you to our reviewers for their insightful comments.

Mya Poe would also like to thank John Cogan, Nancy Poe, and Marvin Poe.

Neal Lerner would also like to thank the support of his family: Tania Baker, Jean Baker, Hannah Baker-Lerner, and Clay Baker-Lerner.

Jennifer Craig would like to thank Chandra, Andrew and Becca, Lois and George, and her friends and family who were so supportive and enthusiastic about this project.

Our research was partially supported by a grant from the Dean of the School of Humanities, Arts, and Social Sciences at MIT, as well as from NSF DUE grant 0341127 for Rigorous Research in Engineering Education. Earlier versions of portions of this book were previously published in *IEEE Transactions on Professional Communication* ("Innovation Across the Curriculum: Three Case Studies in Teaching Science and Engineering Communication," copyright 2008 by IEEE, Jennifer Craig, Mya Poe, and Neal Lerner, reprinted with permission).

Introduction

Engineers who don't write well end up working for engineers who do write well.
—The MIT Writing Across the Curriculum Program

MIT students, by and large, do not love to write. Although that charge might be made against many students at many institutions (and against faculty, for that matter), the science and engineering orientation of MIT undergraduates can often lead them to believe that in their professional careers, the search for engineering solutions or scientific phenomena, not the seemingly tedious process of communicating those findings, will dominate. And while many of our students do have secret writing lives, those activities are centered around what they often describe as "humanities writing," such as fiction, poetry, essays, or informal writing over the Internet. Overall, the idea of writing as a scientist or engineer is somewhat foreign and fairly intimidating to them.

Nevertheless, when students arrive at MIT, they find that they will need to write and speak a great deal and that these activities carry evaluative weight. Most of that activity centers on classes that are designated as communications intensive (CI). Undergraduates are required to take four CI classes: two in a distribution of humanities, arts, and social science classes (CI-H) and two in their majors (CI-M). A 2008 assessment report on students' satisfaction with these classes found that student buy-in was generally good, though seniors placed higher value on the CI-M classes—the writing and speaking they did in their majors—as compared to CI-H classes—the writing and speaking they did outside their majors (MIT 2008). This finding is not particularly surprising, given the exigency of major classes and students' impending start of their post-undergraduate careers. This finding is also in accord with previous longitudinal studies of students' reflections on the importance of writing instruction (Beaufort 2007, Sternglass 1997, Carroll 2002). MIT devotes considerable resources to support CI activities, including a staff of forty-five full- and part-time lecturers who offer supplemental instruction in CI classes. CI teaching and learning activities make a great deal of sense in a climate of increasing demands for science and engineering professionals to be skilled communicators and given what we know about how important writing and speaking are in helping students learn content.

Still, these larger goals potentially mask a whole host of issues that need to be addressed to ensure the success of a communications-intensive requirement—and the administration and instructional components created to support it. This book looks deeply at those issues and at the processes of student learning in a range of CI science and engineering classes. In biology, biological engineering, aeronautics/astronautics, biomedical engineering, and health sciences technology, students are faced with the tasks of writing, speaking, persuading, and collaborating that are modeled closely on professional standards. This work is embedded in the particular cultures of those classes, majors, and professions, adding a host of implicit expectations that shape the evaluation of student work. Finally, student learning occurs in the context of schooling, rather than in the workplace or research lab, and the time constraints, cultures, and demands of MIT strongly shape that experience. Overall, the faculty and instructional staff described in this book are involved in the complex activity of teaching students to write and speak as scientists and engineers and are doing so within specific discourse communities. An improved understanding of the key factors in these contexts can tell us a great deal about what works and what does not, as well as expand our theoretical knowledge of teaching and learning in Writing Across the Curriculum and Writing in the Disciplines programs.

The Call for Communication in Professional Contexts

The classes we describe in this book are not unique to MIT but instead are part of a national trend in engineering and science education toward achieving communication outcomes. In many ways, this trend has been driven by professional organizations that have recognized the importance of good communication skills for engineers. For example, the National Science Foundation (NSF) has long recognized the importance of communication education in the sciences and has encouraged changes in the way that student scientists are educated. NSF supports efforts in science and science writing in various venues, including the Beyond the Classroom professional development program at the University of California Santa Barbara, a "partnership between scientists and science teachers to support the use of technology in science courses for the purpose of enhancing students' scientific investigation and English language skills" (Materials Research Laboratory, n.d.). The American Association for the Advancement of Science (2008) Benchmarks for Science Literacy include as one learning goal to "choose appropriate communication methods for critically analyzing data."

A major force for recognition of the importance of communication skills has also been the influence of accrediting agencies. Since 2000, the Accreditation Board for Engineering and Technology has required that engineering programs demonstrate that their graduates show proficiency in a number of "soft skills," including:

- An ability to design and conduct experiments, as well as to analyze and interpret data,
- An ability to function on multi-disciplinary teams,
- An ability to communicate effectively.

The adoption of these goals has driven a great deal of curricular reform in science and engineering, but their relative abstractness has presented a problem (Paretti and McNair 2008). As Driskill (2000) asks, what does it mean to "communicate effectively"? And what are "communication skills," particularly in light of research on the development of writing abilities and sociological work on the role of communication in the lives of practitioners? On one level, it is tempting to define *communication skills* as grammar or correctness—concrete, easily identifiable elements of good writing, yet writing instructors and their engineering and science colleagues know there is more to good writing than grammar and syntax.

A traditionalist point of view would argue that writing occurs after the science or engineering has been done and that the role of communication is to transmit that information to readers. In this view, scientific or engineering education should focus on the accumulation of factual and procedural knowledge with communication education occurring outside the technical classroom after students acquire technical expertise. Although this appearance of the functionality of communication as a funnel from writer to reader is tempting, it ignores the day-to-day microcommunications that occur in labs and other workplaces. A traditionalist view of communication also does not help students learn the ways that disciplinary values are embedded in the ways of speaking and writing used in science and engineering. For professional scientists and engineers, writing does not occur at the end of research or design; it permeates the research process. As Robert Locke explains, "To be sure, at some stage in scientists' work a writing down (a writing up) occurs; they prepare, finally, written documents, scientific papers or reports, which . . . 'represent' their findings. Yet the specific language in such a paper does not arise *de novo* when scientists come to write; much of it is already present *in posse* in everything they think about their work" (1992, p. 34).

In our experience, working scientists and engineers have little difficulty acknowledging the persuasive aspects involved in their written publications and presentation choices, and they have little problem balancing these rhetorical decisions against the objective persona involved in a science or engineering professional identity. In making decisions, they do not deny the "brute facts" of nature, but they acknowledge that these facts are not science itself (Gross 1990).

For professional scientists and engineers, communication skills include a wide range of abilities, including knowing when and what types of communications to use, how to use evidence that is recognizable and understandable to the audience, how to deploy communication in ways that appeal to a group's sense of itself, and how to work collaboratively to achieve those ends. Such aspects of communication skills are not esoteric qualities. Rather, these skills are teachable through instruction, practice, and feedback. Nevertheless, despite a good deal of professional activity directed toward refining these practices, challenges remain. These include, according to Paretti and McNair (2000), disciplinary faculty not having a foundation in composition and communication practice and theory, institutions unable to create interdisciplinary partnerships between disciplinary faculty and communications specialists, and students not having flexibility in their schedules for communication instruction. To overcome these challenges, our approach at MIT has been to embed CI writing and speaking instruction in the structures of students' classes and, ideally, the cultures of the disciplines. The case studies we describe attest to the challenges and opportunities of this approach.

Communication Instruction and Student Learning at MIT

In our teaching practices and in the research we describe here, we take the approach that communication helps shape scientific and engineering practice by constructing how knowledge is articulated; that is, "scientific knowledge emerges from a nexus of interacting people, agencies, materials, instruments, individual and collective goals and interests, and the histories of all these factors" (McGinn and Roth 1999, p. 15). This approach is deeply indebted to the Writing in the Disciplines (WID) movement, which began in the 1980s along with the Writing Across the Curriculum (WAC) movement as a method to teach students disciplinary ways of communicating in their respective professional fields (Russell 2002; McLeod, Miraglia, Soven, and Thaiss 2001; Pritchard and Honeycutt 2006). First principles of the WID movement are to model the authentic communication genres and processes of professionals as closely as possible within school contexts. More recent innovations in WAC and WID have been toward Communication Across the Curriculum (CxC), which attends not only to written work but also to oral and visual ways of doing communication in the disciplines (Bazerman et al. 2005).

As instructional movements, WAC, WID, and CxC have offered examples of and justification for curricular structures that are now in place in a wide variety of institutions. Less well known is how students learn in these programs, particularly in the context of science and engineering education. Studies point to the need to help students understand the discourses of science and engineering and how to use those discourses to specific ends (McGinn and Roth 1999). Many institutions have created stand-alone writing courses for students in hopes of efficiently improving their academic writing (Leydens and Olds 2007). Although these courses can offer valuable insights into the discursive conventions within a discipline or procedural knowledge about grant submissions or human subjects training, they run the risk of drifting from the actual practices in which science and engineering professionals engage during lab and field research. Moreover, stand-alone courses often neglect oral communication, a key skill for engineers and scientists. The case studies in this book offer a number of innovative oral and written tasks, all based on the exigency of problems to be solved in specific fields, and thus representing authentic opportunities for students to learn to communicate in those fields.

In terms of the context for the classes and students we describe, the integration of writing instruction into science and engineering classes at MIT began in the late nineteenth century (Russell 2002); however, curricular reform to bring about the current communication requirement was the result of alumni feedback gathered in the mid-1990s. While alumni felt that they had received top-notch technical educations, their lack of proficiency with writing and speaking was a significant hurdle to professional success. In response to this feedback, in 1997 MIT initiated multiyear curricular pilot programs involving communication education, and these pilot programs became the basis for the current CI curriculum. In 2000 MIT faculty passed the communication requirement, an institute-wide faculty initiative with the intention to integrate "substantial instruction and practice in writing and speaking into all four years and across all parts of MIT's undergraduate program" (Office of the Communications Requirement 2008).

As we noted previously, the communications requirement requires MIT students to take at least four CI courses in their four years at MIT. Two of these courses must be completed in the major. Such courses emphasize communication in the learning of technical content. Unlike some other universities where such courses might be taught as stand-alone entities or with minimal input from the writing program, MIT's CI courses are taught collaboratively with technical faculty and communications instructors. Each department at MIT develops its own CI courses to reflect the disciplinary needs of its students, some of whom enter industry and some of whom continue to graduate or professional school. This integrated approach develops students' writing and speaking skills in the practice of doing science and engineering.

Our initial goal in designing CI instruction was to work with engineering and science faculty to design meaningful, well-defined assignments, use revision and peer review to improve student writing, develop learning goals, and effectively assess student writing. Fundamental writing and speaking instruction still forms the basis of our collaboration with engineering and science faculty. What has also emerged, however, is a move away from these initial steps to writing and speaking activities that resemble the more advanced challenges of engineering communication that occur in the practice of doing engineering (Carter, Ferzli, and Wiebe 2007). In this way, we have been able to ask not only, "What forms of writing should students be doing?" but also "What activities encourage students to work and to think like professional engineers?" Our particular interest is helping students move from general academic writing or novice approximations of disciplinary writing to internalizing the communication-thinking practices of professional engineers (Leydens and Olds 2007; Bransford, Brown, and Cocking 2000). Thus, our CI classes are tailored to fit the communication practices of the young professionals in the particular discipline in which we are working. Our collaborative work with engineering and science faculty blends our understanding of writing pedagogy with the expectations of the specific discipline.

One might argue that this considerable commitment of resources makes MIT unique in its approach to integrating communication instruction in engineering classes. However, institutions ranging from large state universities such as North Carolina State University to smaller institutions such as Presbyterian College have active communication-intensive programs, albeit using quite different approaches. At some institutions, communication and writing-intensive instruction is offered on a workshop basis to interested faculty. At others, CI instruction has been added to general education requirements and is supported through writing fellows or a writing center. The integrated model we use at MIT is also found at other institutions, although on a smaller but nonetheless quite effective scale. As we describe in the MIT case studies that follow, the realities of working with faculty, staff, and students to help them achieve communication and course goals can be easily applied to a wide variety of settings. The importance of helping students meet the target competencies of professional practice, of teaching effective teamwork and collaboration, and of teaching students to understand and argue with visual data are recognized as widespread needs, particularly in the framework of the Accreditation Board for Engineering and Technology's engineering communication criteria (Shuman, Besterfield-Sacre, and McGourty 2005). We believe that our examples attest to the possibilities and challenges in meeting those needs both inside and outside MIT.

Sociocognitive Theories of Writing, Speaking, and Learning

The teaching and learning reported in this book are informed by the particular curricular needs of our institution and the developing relationship between disciplinary faculty and communication instructors. However, contemporary models for understanding how science and engineering students learn to communicate strongly shape this teaching and the research studies that follow. These models are based on the idea that writing and speaking are essentially sociocognitive acts and that specific communication tasks are intertwined with the social context in which they are situated (Lave 1996, Gee 2000, Prior 2006).

The sociocognitive view is useful for studying student learning because it takes into account human interactions within the contexts in which those interactions occur. In other words, a social model of learning takes into account the many relationships essential to successful study, whether with other students, classroom faculty, or laboratory personnel (McGinn and Roth 1999). A purely cognitive view of learning, while useful, does not consider the social or rhetorical purposes that writing serves in communities. Science and engineering, like other communities, have their own ways of getting things done, their own internal disputes, and their own ways of inculcating new members into the community. Teaching writing without paying attention to these dimensions of learning to become a scientist or engineer risks disconnecting writing from the practice of doing science and engineering and further relegating writing as a remedial task that is to be acquired outside science and engineering. Research has long shown that decontextualized approaches to teaching writing, particularly disciplinary forms of writing, do not yield the kind of learning that comes with learning to write within specific contexts and performing the communication activities typically done within those contexts (Smith, Cheville, and Hilloks 2006).

This socially situated view of learning encompasses four factors present in school-based contexts:

Faculty The experiences, expectations, beliefs, and skills that faculty and mentors bring to the learning context mediate students' learning.

Students The experiences, expectations, beliefs, and skills that students bring to the learning context shape students' readiness for learning, as well as the outcomes of their learning.

Learning contexts The contexts for learning to communicate—whether classroom, laboratory, research group, or peer group—all present different or distributed opportunities for learning that cannot be isolated solely in classroom settings.

Communication tasks and processes The particular communication tasks (processes, practices, and genres) that are undertaken are shaped by the institutional setting as well as the values and functions of the larger professional community.

Each of these areas is represented by considerable bodies of research, which we next review to indicate the theoretical grounding for the research that follows and the foundation for the CI program at MIT.

The Role of Faculty, Instructors, and Mentors

One way to theorize sociocognitive factors in learning is to view the learning context for science and engineering students as one in which they are novice learners in a novice/expert system. The two dimensions to this theory largely draw from the work of Soviet theorist Lev Vygotsky (1962). One is the idea of the *zone of proximal development*, or the achievement that a novice might accomplish with the guidance of a more knowledgeable peer or mentor. Key to this instruction is the need to provide scaffolding or the support structures that introduce novice learners to professional practices and enable them to perceive and undertake communication tasks more effectively than they would be able to were they working alone (Wood, Bruner, and Ross 1976). Also essential is that the expert or mentor in the system makes visible the thinking and problem-solving processes underlying her performance. The second contribution based on Vygotsky's theories is the concept of learning as a *cognitive apprenticeship* (Collins, Brown, and Newman 1989). This model asserts that learning comes from gradual and guided participation in the communication activities essential to the system (Lave and Wenger 1991). For science and engineering students, this means learning the structures of professional practice in which communication occurs, that is, attending to the wider network of situations in which they need to write and speak.

Ideas of apprenticeship offer attractive ways of envisioning a teaching-learning environment that is in contrast to traditional schooling (Collins, Brown, and Newman 1989; Rogoff 1995). Apprenticeships "characterized learning before there were schools, from learning one's language to learning how to run an empire" (Collins, Brown, and Newman 1989, p. 491). For Rogoff (1995), apprenticeship is not merely one master teaching his or her craft to an eager novice. Instead, "apprenticeship as a concept goes beyond expert-novice dyads; it focuses on a system of interpersonal involvements and arrangements in which people engage in culturally organized activity in which apprentices become more responsible participants" (p. 4). Collins, Brown, and Newman use the term *cognitive apprenticeship* in contrast to traditional apprenticeship in order to highlight two key differences: (1) apprenticeship in school settings emphasizes an expert's processes of problem solving in which "conceptual and factual knowledge are exemplified and situated in the contexts of their use" (p. 457), and (2) cognitive apprenticeship "refers to the focus of the learning-through-guided-experience on cognitive and metacognitive, rather than physical skills and processes" (p. 457). In other words, the intellectual work of schooling is the subject matter of cognitive apprenticeship, and it is through the activities of observation, coaching, and independent practice that students develop as successful learners.

In science and engineering education, mentoring and cognitive apprenticeship models are particularly prominent. In the case studies in this book, we show mentoring in a variety of dimensions, which include the kind of professional standards or competencies asserted by a professional's comments on a student draft or oral presentation, as well as more direct modeling of scientific thinking and problem solving.

What Students Bring to Learning

The connection between identity formation and learning is a central concept to social theories of learning, in which identity is socially constructed through a person's interactions with others, with knowledge, and with the physical and symbolic elements that he or she uses to communicate (Gee 2004). Central to identity formation are the motivations, beliefs, and attitudes that students bring to the learning context, and uncovering these affective dimensions is essential to both research that attempts to understand the factor that may limit or enhance newcomers acquisition of new skills (Blakeslee 1997) and to design interventions that capitalize on students' growing sense of identities as professionals. The research we report on in this book relies largely on interviews with students to understand these developing identities.

For students writing about their scientific research, identity also comes into play in terms of knowledge-making processes. Leydens (2008), in a study of engineers at various professional stages (including data from when some were still students), found that participants' conception of their role in the knowledge-making process—whether as a relatively static conveyor of relatively static knowledge or active shaper of meaning through rhetorical prowess—marked the path from novice to expert communicator. In Leyden's words, the most experienced engineering writers "enact identities as confident change agents" (p. 254). Identity formation—particularly a "discursive identity" or that formed by engaging in particular communication tasks (Brown, Reveles, and Kelly 2005)—is particular important for the students we describe in this book.

Contexts and Processes for Learning

Another element essential to sociocognitive theories of learning is the role of the context in which that learning is taking place. Whether students are learning in the classroom, the lab, or the field, each site socially situates learning activities, and a challenge for instruction becomes how to transfer student learning from those contexts to new and unfamiliar ones. Research on science and engineering students making the transition from school to workplace contexts demonstrates the ways that individual contexts strongly determine the means, forms, and success of communicative acts (Freedman and Adam 1996; Freedman, Adam, and Smart 1994; Beaufort 2007; Leydens 2008).

One way to think of science and engineering contexts in regard to communication are as discourse communities. In a simple sense, discourse communities comprise individuals who share certain language-using practices (Bizzell 1992). (For elaboration, see Swales 1990 and Porter 1992.) Typically when we talk about professional contexts for language use, we talk about discourse communities. Tissue engineers, for example, share a common way of talking about the human body, even if they do not always agree. Discourse communities need not be rooted at a physical location, but they do need a context for their practices to occur. Context, in this sense, is a place for the accumulation of wisdom, a place for members to air their grievances, and a place for new members to be initiated into the style, norms, and ideologies of the community.

When talking about school, discourse communities might be better labeled as "learning communities." Learning communities are, Riel writes, a "way of knowing, a set of practices and shared value of the knowledge that comes from these practices" (1998, p. 1). A learning community might be more specifically described as a setting in which the community is organized rather than disciplined; characterized by collaboration rather than competition; focused on knowledge construction rather than knowledge delivery from one central source; student centered rather than teacher centered; interdependent rather than strictly independent. Instead of expertise flowing from the teacher to many students, expertise flows in many directions. Community members are recognized for what they know and can do, while leaders are people who inspire others to work toward common goals (Riel 1998). Smith et al. (2005) describe this kind of learning community as "engaged."

Anchoring instruction in authentic activities from the science and engineering professional world also calls for students to engage in steady collaboration and requires them to develop team skills. Collaborative communication within science and engineering education is based in the activities that naturally occur in design and research. Although the academic environment can never completely mimic the professional workplace, there is (or can be) enough verisimilitude within most design and research projects to introduce students to the authentic requirements of collaborative communication and teamwork.

Cooperative learning is often bundled with collaborative learning, and although these two approaches are similar, they have different historical roots. Collaborative approaches emphasize student interaction rather than solitary activities. Cooperative learning has a similar emphasis and in addition emphasizes cooperation and mutuality over competition (Prince 2004). The efficacy attributed to cooperative learning comes from several assertions. Cooperative learning can reduce the unproductive competitiveness that individually focused instruction can sometimes produce. By valuing the success of the group, students help one another to meet the group goal. Cooperative learning groups can create a supportive and collegial atmosphere in which to learn, thereby enhancing learning. Cooperative learning also allows groups of students to approach larger and more complex problems because the groups can offer a range of skill sets and increased critical thinking capacity and energy for problem solving (Springer, Stanne, and Donovan 1999).

In terms of team skills, when cooperative learning is structured around a realistic task (design, research, writing, presenting), it also offers a "natural environment" in which team skills can be practiced and developed (Prince 2004). In fact, Burrell and Colton (1999) argue that team activities should be "highly contextualized to that point that [they] are an inseparable part of what is normally done in the course" (p. 1). As student and faculty surveys and interviews have articulated, the experience of teamwork is the most effective way to learn how to work in a team. Certainly a modicum of explicit instruction and a good deal of reflection and assessment help refine and expand team skills, but the cooperative experience is fundamental to learning.

While there may be challenges in measuring the efficacy of certain models, researchers agree that the data are conclusive: active learning and, more specifically, collaborative and cooperative learning are effective in undergraduate science, mathematics, and engineering and technology courses and programs (Prince 2004; Springer et al. 1999; Smith 1995).

Communication Tasks and Forms

A feature of discourse communities is their use of shared genres to advance communication goals. One of the hallmarks of membership in a discourse community is fluency with the forms, or genres, of writing and speaking used in that community. However, as North American genre theorists have suggested, genres are not merely a collection of conventions of written or spoken discourse. Genres, or typified rhetorical reactions to recurrent situations (Miller 1979), allow members of a community to interact in a way that signals their membership in that community. Embedded in these forms are linguistic markers to a community's ways of knowing and arguing, and to its values (Driver, Newton, and Osborne 1997). Readers pick up on these markers and make assumptions based on them "about the text's purposes, its subject matter, its writer, and its expected reader" (Devitt 2004, p. 12).

North American genre theory has played a particularly strong role in research on students' learning in science and engineering (see, e.g., Luzon 2005; Walker 1999; Russell 1997; Freedman, Adam, and Smart 1994; Artemeva 2005; Artemeva, Logie, and St-Martin 1999). Key features of this theoretical perspective are that communication forms, such as a laboratory report, are not static but instead are shaped by the contexts in which they are produced and the social exigency of that production. Unfortunately, many science and engineering classrooms and how-to guides present scientific texts as relatively static products, seemingly codified through years of repetition. In other words, students learn "what" but rarely are offered the "why" or the knowledge production possibilities essential to a genre approach. For example, the long history of the school-based laboratory report as a "plug-and-chug" format or regurgitation of content in static forms ignores the critical role of the scientific report within the discourse communities of scientists. From a genre studies perspective, the scientific report shapes and is shaped by the needs of writers and readers; its production is a meaning-making activity occurring for particular social needs, and its more stable formats are a result of patterns of those needs (Bazerman 1988).

Another key feature of research on genre has been the observation that genres do not work in isolation. They travel together in sets and even operate in entire systems, all supporting a human activity (Bazerman 2004). What this means is that students must become knowledgeable not just about a single genre, such as a research article, but about the interrelationships among official genres in systems (e.g., letter of intent, grant proposal, supplemental reviewer material, reviews) as well as the unofficial or supporting genres (e.g., calls to program officer, e-mails to collaborators) (Berkenkotter and Huckin 1995; Spinuzzi 2003).

In the class and lab contexts we offer in this book, students were writing and presenting a variety of forms: design reports, PowerPoint presentations, research articles, laboratory reports, and business plans, among others. Certainly some of these forms are more codified than others (and thus in some, students had more leeway to veer from standard formats), but all forms are products of social action, and from a genre studies perspective, it is essential to recognize that students ideally become players in that production process.

Studies of Writing and Speaking in Sociocognitive Environments

In addition to the theoretical context for our research, the case studies that we present in this book are informed by research on students' learning to write and speak in disciplinary contexts, particularly in science and engineering. The methods, findings, and analysis from these studies provide a disciplinary context for our work, as well as the research gap (Swales 1990) that we felt needed to be addressed.

Studies of the link between communication instruction and learning have produced mixed results, in some showing a positive impact on student learning while others showing inconclusive results (Bangert-Downs, Hurley, and Wilkinson 2004; Oschsner and Fowler 2004; Klein 1999). Students who engage only in traditional schooling writing activities such as note taking evidence little change in learning (Tynjala, Mason, and Loonka, 2001; Langer and Applebee 1987). Students who participate in writing assignments that model those genres and inquiry activities that professionals in their field use learn communication within the social practices of their disciplinary communities and develop new verbal abilities as the result of that socialization (Keys 1999, Luzon 2005, Freedman et al. 2004, Carter et al. 2007).

In research specific to students' writing in science and engineering, early qualitative studies include Herrington's (1985) study of chemical engineering students in two classes; Walvoord and McCarthy's (1990) study of students in four disciplines, one being biology (see also McCarthy 1987, which included a student taking a cell biology course); and Haas's (1994) study of one undergraduate learning biology. Herrington's study, which relied on survey data as well as interviews and analysis of written texts, showed that students could be introduced to the social roles of a discipline through writing but that they may have difficulty in shifting from a school to a professional context. Her study also showed that each chemical engineering class represented its own discourse community: "Even within one discipline, chemical engineering, different courses may represent distinct forums where different issues are addressed, different lines of reasoning used, different writer and audience roles assumed, and different social purposes served by writing" (1985, p. 354). The main insight from Herrington's study was that a monolithic sense of academic disciplines as homogeneous communities was overstated, even in school contexts (a claim supported by sociologists).

Walvoord and McCarthy's study, which included think-aloud protocols, interviews, and analysis of student writing, also showed a link between student writing and the social role students were expected to enact: "Students experienced difficulties not only in adopting the role of scientist, but also in performing it appropriately—that is, using the scientific method and writing their reports in the appropriate format" (p. 226). Winsor's (1996, 2003) studies of two groups of engineering students over five years have provided specific insights related to the socialization of engineering students into the profession through the written genres and textual negotiations they encountered in a cooperative learning program and in the workplace. Finally, Beaufort (2007) tracked one student, Tim, from the writing he did as an undergraduate with a double major in history and engineering to the writing he was doing on the job as an engineer after graduation. Beaufort found that Tim's education gave some measure of preparation for the challenges he faced writing on the job, but that the learning curve was still steep.

These studies show us that writing development is not linear. They also tell us that the development of writing ability is related to the other developmental changes going on in students' lives. And finally, these studies tell us that students learn to write through instruction and practice. The ethnographic studies on writing development in the disciplines have been insightful because they use the disciplinary community's own definitions of effective communication rather than externally imposed measures. Additionally, such methods are best suited to capturing the full range of contextual factors that affect students' learning of communication skills (see also Patton and Nagelhout 2004, Geller 2005).

The Structure of Learning to Communicate in Science and Engineering

The chapters that follow describe students as they learn to write and speak in five different classes and disciplines. Each chapter has a particular frame: the role of identity for students in biology, the importance of authentic communication tasks for students in biological engineering, finding a research niche in health sciences technology, learning how to persuade with visual data in biomedical engineering, and the relationship of collaborative communication and teamwork in aeronautic/astronautic engineering. As our teaching of writing and speaking has been integrated into disciplinary subjects, these topics have emerged not simply as concepts but more specifically as practices integral to the instructional goals of CI work in the disciplines. In the classes described, we illustrate the implementation and refinement of these practices. From our teaching in these class contexts, we designed research to explore the practices that we believe have been largely underexplored in terms of previous research. Our research conducted over one or two semesters employed interviews with and surveys of students and instructional staff, observations of class and laboratory sessions, and analysis of students' written, visual, and oral work, as well as assignment sheets and departmental descriptions. (Appendixes A and B contain detailed descriptions of the methodology, survey instruments, and interview questions.) Student communication tasks included scientific articles, grant proposals, design proposals and reports, experimental reports, oral design reviews, PowerPoint presentations, business plans, and editorials. Based on these data sources, each chapter explores how identity, authenticity, persuasion, and teamwork are embodied, practiced, acknowledged, and resisted in actual classrooms.

The courses that we profile are not representative of every discipline at MIT. Instead, they represent courses that we have identified as having developed highly effective and efficient CI instruction methods—current "best practices." Each chapter also presents courses that might follow from one another so readers can see the learning trajectory of students as they move from one CI experience to the next. In the chapters that follow, student names have been changed, and particular demographic criteria have been modified or edited to protect student confidentiality. We can report that our limited sample of students and teachers represents a wide spectrum of gender, lifestyle, age, and ethnicity. The seventeen students in the seven classes studied in this book were consistent with the diversity found at MIT, where nearly half of the undergraduates are women and nearly half are from U.S. minority groups (*MIT Facts* 2007). Our student group had both native-born and international students. Also, a number of the faculty and teaching assistants in this study were women, representing not only the contemporary face of engineering and science but also MIT's effort to enhance ethnic and gender diversity across the campus. Although we did not specifically address these identities in our chapters, such identities are certainly important to students' educational experiences. It would be fruitful in subsequent research to understand more how these identities shape student learning. While the WAC and WID research has considered gendered identity (Wolfe 2005), it has been virtually silent on issues of racial/ethnic identity and sexual orientation (Anson forthcoming).

In chapter 1, we examine the ways in which undergraduates in biology take on the discursive identities of professional scientists as they work to turn scientific findings into a research article. The development of this professional identity guides the case studies in this chapter as we explore the following topics:

- What are students' challenges and opportunities as they face the dominant writing task of scientists: the scientific research article?
- How do students' identities as science students and neophyte scientists shape teaching and learning in a molecular biology laboratory class?

Key to this chapter are the developmental steps students take as they learn that scientific communication is more than efficient transfer of knowledge and includes a wide range of rhetorical concerns.

In chapter 2, biological engineering majors take more fully developed scientific identities and apply them to the range of authentic tasks that biological engineers encounter. In this chapter, we ask,

- What is the relationship between authentic writing tasks and students' development of scientific discursive identities?
- What are students' challenges and opportunities as they face multiple writing tasks of biological engineers?

The students featured in chapter 2 are more advanced in their academic careers and have a clearer sense of their futures than students in chapter 1. However, this clarity was not always an ally in their completion of the writing they faced, as the need to develop facility in and understand the importance of a variety of genres was a consistent challenge.

In the chapter 3, we examine how grant writing brings forth certain tensions as graduate students in health sciences technology learn to define and motivate a research agenda. In this chapter, we focus on three questions:

- How do students learn to motivate their scientific ideas in an organized fashion that appears significant to other scientists?
- What is the role of a student's mentor in learning to stake out a research territory?
- How do authentic writing activities like grant writing contribute to student learning?

Key to this chapter is the pivotal role of a graduate student's mentor in the learning process and the importance of authentic peer review.

In chapter 4, we examine the ways in which undergraduates in biomedical engineering learn to use data as evidence. Specifically, we look at the ways that students are taught to argue like scientists and explore the following questions:

- How do students learn the persuasive devices that professional scientists use when communicating data to other scientists?
- What challenges do students encounter in learning how to use visual evidence in scientific communication?
- What role does faculty feedback play in the development of this professional skill?

In chapter 5, we turn to the challenges of collaborative communication in aeronautical/astronautical engineering and the ways in which students learn to communicate collaboratively and develop the team skills so central to collaboration. We explore three questions:

- How do students efficiently and effectively learn to write and present collaboratively?
- What specific team skills are central to this task?
- How are those team skills best learned? By instruction? By mentoring? By experience?

Each chapter begins with an introduction that contextualizes the studies in that chapter and is followed by a description of the goals for student learning and communication activities. We then present case studies of individual students and overall findings from this research of undergraduate and graduate students as they learn to write and speak in a variety of science and engineering classes at MIT. In each chapter, we refer to students and teaching assistants by a first-name pseudonym to ensure confidentiality. We refer to faculty by their full name at first use and then last name subsequently. They have given us permission to use their names in this book.

We conclude the book with implications for pedagogy. In terms of those implications for any department wishing to design its own program, it is important to note that the structures of the CI courses presented in this book are quite different from one another. In this book, we present three types of CI courses. First are whole, or integrated, models in which the CI component is integrated entirely with the technical content of the class and works independently of other CI classes (e.g., chapters 2, 3, 4, and 5). Second are "stretch" models in which the CI curriculum dovetails with the CI curriculum in a subsequent class (e.g., chapter 5). Third are what we call "sidecar" classes in which the CI curriculum is independent of the technical course curriculum. These include tutorial models in which students meet once a week to discuss drafts of their research findings (e.g., chapter 1). The design of each course reflects a combination of faculty input, departmental resources and commitment, and goals for student learning.

It is important to note that it is not our intent to create a divide between what is possible at a resource-rich institution such as MIT and other institutions not similarly fortunate. If anything, the questions we raise in these case studies and the limits of our instruction are applicable to many institutions struggling to create meaningful educational opportunities in engineering communication.

Our research and the case studies we present raise other widely applicable questions: What does it mean for educational practice if professional communication competencies and tasks are the goals? How can engineering students move from mere display of data to making skilled visual arguments based on those data? How can students and technical faculty best create the conditions for students to learn to be skilled team members? By no means have we figured out the complete answers to those questions, but we hope that the case studies we present show some potential paths to finding those answers

Our inquiry into the teaching of communication at MIT also represents "teacher-research" or "action-research" (Cochran-Smith and Lytle 1993) meant both to improve practice and to broaden our knowledge of what it means to learn to communicate as a scientist and engineer. Thus, the sustained research in which we continue to engage through surveys, focus groups, individual interviews, and analyses of students' work, all in the context of contemporary theories of teaching and learning, are essential activities for communication professionals in any setting.

The case studies we present here are not the final word on the redesign of communication instruction in engineering education. Our intent instead has been to highlight the ways that a commitment to teaching communication within disciplinary frameworks at MIT has brought to the four key aspects that require attention: identity, authenticity, argumentation, and teamwork and collaboration. Each of these aspects is present in some degree to all of our CI classes, and each reveals the opportunities and complications for designing communication instruction.

1 First Steps in Writing a Scientific Identity

Throughout this book, the writing and speaking tasks that science and engineering students engage in are largely modeled on professional tasks and genres, including research articles, poster presentations, and grant proposals. The use of professional-like tasks calls for students to assume identities as scientists or engineers as they engage in these apprenticeship activities. The development of this professional identity guides the case studies in this chapter as we explore the following questions:

- What are students' challenges and opportunities as they face the dominant writing task of scientists: the scientific research article?
- How do students' identities as science students and neophyte scientists shape teaching and learning in a molecular biology laboratory class?

As we noted in the Introduction as we reviewed the social theories of learning that inform this book, identity is a key concept for students as they learn to write and speak in science and engineering classes. Literacy theorist James Gee makes the connection between identity and learning as follows:

> Knowledge and intelligence reside not solely in heads, but, rather, are distributed across the social practices (including language practices) and the various tools, technologies and semiotic systems that a given "community of practice" uses in order to carry out its characteristic activities. . . . Knowing is a matter of being able to participate centrally in practice, and learning is a matter of changing patterns of participation (with concomitant changes in identity). (2000, p. 181)

In terms of students learning to write science, participation in the communities of practice of scientists represents changing patterns of participation, which in turn potentially alters students' sense of who they are or will be as scientists. This notion of students as novice professionals learning to write and speak successfully in their chosen fields leads to the need for instruction from professionals in those fields, a strong feature of the courses and students profiled in this book. Learning in these settings is thus a form of apprenticeship, another key term for social theories of learning and one explored in this chapter.

Nevertheless, while the work of professional scientists and engineers is shaping teaching and learning, communication activities are still occurring within the aegis of the classroom or school-based laboratory. In many of these rhetorical situations, the classroom teacher is the ultimate reader and evaluator of students' texts, and the strongest identities that students assert are their identities as students. Dannels (2000) found that for student groups engaging in real-world tasks in mechanical engineering, the context of the classroom itself, rather than the professional goal, strongly determined their actions. As was true in Dannels's study, for many of the students profiled in this chapter and in this book overall, the powerful influences of schooling were consistent factors, and it would be naive to ignore them (see also Freedman, Adam, and Smart 1994; Freedman and Adam 1996).

For many MIT students, outside-of-class experiences have provided strong technical backgrounds in experimental science. Often these are highly valued experiences: working in research laboratories as interns or during summer projects and competing for science prizes and competitions, for example. However, within those activities, students' roles rarely include writing up that research for publication or communicating what they have learned. In a sense, students' "discursive identities" (Brown, Reveles, and Kelly 2005) or their sense of themselves as scientists (or science students, for that matter) as expressed through their writing and speaking about science are barely formed. As the case studies that follow show, most students are only beginning to assert identities as scientists. Those identities are in flux as they sample majors, have their first significant laboratory experiences, and balance the intellectual work required in that context with the time they decide to make available as busy students.

The development of a scientific identity belies notions of learning to write as simply a matter of following a protocol. Instead, students strive to convey often messy scientific results and distill meaning from those results, while at the same time working within the IMRD (introduction, methods, results, discussion) form valued by scientific professionals. As shown by the Biology Department students profiled in this chapter, writing tasks in biology are also fraught with the complications that always accompany writing tasks: lack of clarity on the tasks themselves, the influence of previous experiences, allocation of time and attention, the mixed messages they might receive for the goals of the assignment, or the varied audiences for whom they are writing. Thus, in addition to the usual rhetorical complexity of scientific writing tasks, the rhetorical situation for school-based laboratory reports or other writing tasks is affected by elements inherent to schooling and students' identities *as students*, particularly the final evaluation or grade that will be assigned to the report. For better or worse (and usually for worse), MIT students can be strongly driven by grades, both as motivators and as indicators of how much time to allocate to any task. These optimizing behaviors add a constraint that is yet another element among a host of social forces that shape the teaching and learning taking place.

Learning to Write in Introduction to Experimental Biology and Communication

The biology class profiled in this chapter is particularly apt for studying the development of students' identities in apprentice-like settings. Biology is a well-established field and major at MIT (though the laboratory class itself has students exploring relatively new technologies), and instruction in its communication-intensive (CI) courses is geared to teaching students to write up their laboratory work as professional biologists would. This assertion of identity, however, is complicated in the class presented here, Introduction to Experimental Biology and Communication (hereafter referred to as Experimental Biology), as few students are headed toward research careers in which they will need to write research articles. Because Experimental Biology fulfills a premed requirement and an MIT laboratory requirement and has relatively large numbers, the enrolled students have fairly diverse majors. During the semester studied, of the ninety-one students enrolled at the start, slightly more than a third, or thirty-two, were listed as biology majors, while twenty-eight were chemical engineering majors, with additional multiple representatives from chemistry, physics, aeronautic/astronautic engineering, brain and cognitive science, nuclear engineering, mathematics, and mechanical engineering. For some of these students, future careers might encompass bench research and the need to write up scientific findings, but many more students will engage in the myriad writing tasks of the various science, medical, and engineering professions they will pursue. Thus, engaging in authentic tasks to develop an identity that maps to students' professional pursuits is a relatively difficult target.

Nevertheless, the writing that students do in Experimental Biology has been identified by the Biology Department as a key component of the course, and particular resources have been focused as a result. For example, when MIT adopted a CI course requirement, Experimental Biology was a natural fit as a CI course, given that students had been writing up their experimental work for many years. However, under the new requirement, the class went from a twelve-credit class to an eighteen-credit class to accommodate and acknowledge the additional work students would be doing. More important, rather than fold writing instruction into the existing course routines, a series of writing workshops was created in conjunction with the Experimental Biology laboratory instructor at the time and two writing instructors from the Writing Across the Curriculum Program, one of whom, Marilee Ogren, is the instructor of the writing section described here. These writing workshops are in addition to the two hours per week of lecture, eight to ten hours per week of lab, and one to two hours per week of recitation students are expected to attend in Experimental Biology.

Students are assigned to these instructional workshops, known as Scientific Communication, or SciComm, in groups ranging in size from six to twenty, and these whole groups meet for two hours five times over the course of a semester. Students also meet an additional three to five times for one-to-one or small group meetings with their SciComm instructor to workshop writing in progress. SciComm is 25 percent of students' overall Experimental Biology grade.

In terms of the learning objectives of SciComm and their relationship to the development of a scientific identity, students were presented with the following goals during the semester under study (spring 2008):

At the conclusion of this class, students will be able to:

1. Understand the seven components (title, abstract, introduction, methods, results, discussion/conclusion, tables/figures) of a laboratory research paper.
2. Understand the writing process and its application to scientific writing.
3. Understand the importance of communicating in writing as a scientist.
4. Apply an understanding of scientific writing to their subsequent independent research. (SciComm Syllabus 2008)

Thus, in this genre-based and process-oriented instruction (as indicated by the first two goals), student outcomes are geared toward professional roles or identities (goal 3), learning the writing behaviors of professional scientists (goal 2), and mastering scientific writing that would then be applied to new authentic contexts (goal 4).

Instruction during the five whole-group SciComm meetings is focused on learning a professional genre—the IMRD structure of the research article: introduction, materials and methods, results and figures, discussion and conclusions, titles and abstracts. The focus on this particular genre is a rhetorical one; in other words, rather than having students learn what material goes in each section, SciComm offers them the opportunity to learn why each section of a research article has a particular shape and how professional scientists make deliberate discursive choices based on findings, interpretation of those findings, intended publication venue, and potential readers' reception of the overall ideas and approach. Few SciComm students have previous experience writing in this way, though most are quite familiar—and dissatisfied—with "plug-and-chug" laboratory reports as often taught in high school science. SciComm instructor Ogren says that SciComm students "have no appreciation for what are the components of a scientific paper. That's all new. And no one has taken the time to make that explicit, even if they've been involved in a publication before. And so they come with a huge dearth of knowledge about the mechanics and the principles of writing a peer-reviewed research article. But that's not really a weakness, I mean, that's why they're here, to be educated about those things."

Students' lack of experience with writing scientific research articles also results in a lack of knowledge about the importance of writing and revision to the discursive practices of scientists; thus, drafting and revision are central activities in SciComm in the hope that students' development of scientific discursive identities will include not merely knowledge of form, but knowledge of the rhetorical requirements of that form and of the writing behaviors common to professional scientists.

In terms of the content of students' SciComm research articles, for the semester presented here (spring 2008), students wrote up the laboratory work they did during the recombinant DNA and biochemistry module of Experimental Biology and were expected to offer that report in the form that mirrored the authentic task of a publishable research article. Students were investigating how mutations to the archaeabacterium *Pyrococcus furiosus* (*Pfu*) would affect the performance of this DNA polymerase as compared to the nonmutated or *wildtype Pfu*. DNA polymerase is a vital enzyme in the process of DNA replication as it both enables the replication process to occur and ensures the accuracy of that process. Thermostable polymerases such as *Pfu* are key components in the lab-based process by which DNA is rapidly multiplied, polymerase chain reaction (PCR), which occurs at high temperatures and thus requires DNA polymerases that can withstand such conditions (*Pfu* and other such polymerases were originally discovered in undersea heat vents). Nevertheless, a great deal is unknown about the relationship between the structure of DNA polymerase and its function, and this line of experimentation has the goal of shedding more light on this relationship as well as improving PCR performance with a mutant version of *Pfu*. In addition to investigating this problem, students were learning molecular biology techniques such as site-directed mutagenesis, DNA purification and restriction digestion, recombinant protein expression and purification, a PCR assay, and a forward genetic screen (*7.02 Manual*, Fall 2007).

As far as SciComm students' emerging sense of their identities as biologists or scientists, the results from this study were decidedly mixed. Based on whole-class surveys administered at the beginning and end of the semester, students' overall sense of what it means to write like a biologist was primarily focused on matters of format and structure rather than on rhetorical knowledge or meaning making. In a sense, the instructional activity and focus on the parts of the research paper resulted in students' focus on those parts (understanding the elements of the research article—goal 1 from the course syllabus) rather than on the relationship among those parts, the process involved in creating them, and the larger meaning making that scientists engage in (goals 2 and 3). For the four case study participants, however, these results are more nuanced, and certainly learning to write like a scientist did occur in different measures. For several students uncertain about just what they would do beyond Experimental Biology and beyond MIT, this uncertainty over professional outcomes made for uncertainty over the lessons learned by writing in Experimental Biology. Overall, for many of these students, the specific requirements of the research article (and, more specifically, the research article being assigned and graded in this class) were what was learned—not necessarily a bad outcome, but possibly a limited one.

Learning from SciComm: Survey-Based Results

To understand students' development of discursive identities in SciComm, one data collection technique was to survey students at the beginning and end of the semester in terms of their previous experiences with scientific writing, their knowledge of the components of a research article, and what they felt to be the purposes for scientific writing (see appendix B for these surveys). These questions followed the work of Ellis and colleagues on undergraduate learning of science (Ellis 2004; Ellis, Taylor, and Drury 2006) in which it was found that students with more sophisticated notions of the purpose of writing science (e.g., to learn the science itself, to engage in rhetorical practices) had higher achievement overall in their first-year biology course than students whose conceptions of what they learned by writing in their science courses was focused on mastering a specific form or process. In other words, a more professional scientific identity as seen in a more nuanced understanding of the role of scientific writing and of the relationship between writing science and learning science had a strong relationship with students' overall achievement.

Thus, in the initial SciComm survey, students were asked to list their experiences with scientific-technical writing and then were asked, "When you wrote these scientific documents (e.g., research articles, lab reports, technical reports), what did you feel you were learning?" In the end-of-term survey students were asked, "What do you feel was the most important thing you learned in SciComm?" and "When you wrote your SciComm Writing Project, what did you feel you were learning (e.g., format of a research article, science of Pfu, experimental methodology)?" The assumption was that students' conception of learning through writing about science would become

Table 1.1
Forms of scientific writing students reported writing before SciComm

Lab reports	Research articles and technical papers	Scientific posters	No experiences
80% (55)	35% (24)	3% (2)	12% (8)

Note: Sixty-nine students responded.

Table 1.2
Students' start- and end-of-semester survey responses to what they feel they learned by writing science

When you were writing up your science, what did you feel you were learning?	Learned clear and concise communication/Learned the format of the research article	Learned about the science
Start of semester (62 total responses)	68% (42)	21% (13)
End of semester (41 total responses)	88% (36)	24% (10)

more sophisticated—or more professional—by the end of the term; in other words, their identities as biologists would be more developed.

In response to the initial question about experiences with scientific writing, 80 percent of the total students surveyed indicated limited experiences, citing "lab reports" as most common and often describing these tasks as "high school lab reports" or "just lab reports." As shown in table 1.1 far fewer engaged in more authentic scientific writing tasks such as research articles, technical papers, or posters. In addition, 12 percent reported no previous experiences with scientific writing.

In terms of what students felt they had learned from engaging in these tasks (see table 1.2), a majority (68 percent) described an understanding of scientific writing as "clear and concise" communication as their learning outcome or described what they were learning in terms such as "learning how to communicate effectively, more concisely," or "I felt I was learning the basic methods of how to write a logical report with all the main components," or "how to present data and analysis clearly." A much smaller percentage of students, 21 percent, described learning about the science as an outcome with answers such as, "I used it as a way to pull together everything that we did in the lab. It was a good way to look back at the experiment as a whole," or "I felt I was mostly learning about the content of the material (i.e., gaining a deeper understanding of the scientific research material) and how to put my findings into words." And several students saw very little learning as a result of their previous scientific writing experiences, remarking, "I didn't feel like I was learning. I felt like I was regurgitating old information—my writing did not reflect my own ideas as much," or "Honestly, nothing—I was more focused on transmitting/communicating than in engaging in any sort of introspective process." Thus, most students—as expected given their limited experiences—described scientific writing as a process of translation rather than a method of learning science or engaging in persuasive or rhetorical activities. In other words, they saw themselves taking some technical finding or problem and using clear and concise language and specific formatting to describe it.

Given the large majority of students who initially saw scientific writing as mostly conveying scientific findings in clear and concise language rather than a rhetorical practice and given the goals of SciComm to reveal the deeply rhetorical nature of writing science as part of students' development of scientific identities, we would have expected students' end-of-term surveys to reflect this developing sophistication. However, end-of-term results do not support this conclusion. When asked in the final survey, "What do you feel was the most important thing you learned in SciComm?" 88 percent of the forty-two students who responded reported a learning outcome that was usually expressed as "how to write a scientific report" or "the correct format for a biological paper." Only 12 percent of the total reported learning something about their writing or revision processes, such as the response, "Learning some of the weaknesses in my writing and how to improve on them (with the rewrite)."

When asked at the end of the semester, "When you wrote your SciComm Writing Project, what did you feel you were learning (e.g., format of a research article, science of Pfu, experimental methodology)?" students' responses as shown in table 1.2 were quite similar to the start-of-semester survey: the vast majority of students (88 percent) described outcomes closely matching the first question, such as "format of a research article" or "format—what's included and how that's stated/organized." Still, some students did express an outcome that indicated a more sophisticated discursive scientific identity, even if that outcome was paired with learning structural components. Nearly a quarter of the total thought that they primarily learned about the science, indicating that the writing "not only made me think/reconsider the experiment and the ideas involved but also made me check the organization and style of my writing" or "forced me to go back and make sure I understood the experiment well enough to write about it, so I also learned about the science."

Still, if most students' take-home message from SciComm was primarily about translation of scientific knowledge from specialized content to clear and concise language, the lessons for developing discursive identities of scientists are muted. Surveys, of course, are often relatively blunt instruments for exploring the processes of student learning. Interviews with and analysis of writing produced by four SciComm students described next provide more nuanced findings.

The four case studies that follow feature students in SciComm sections taught by Marilee Ogren, who by this semester had been teaching SciComm for twelve consecutive semesters. With a Ph.D. in neurobiology and extensive experience as a scientific writer and teacher, Ogren sought to convey her professional values for scientific writing, particularly the idea that scientists value writing that is "clear and concise." In large measure, she embodied a professional writer's identity that she hoped would provide a model for her SciComm students. In an interview, she told of a two-year stint with the *New England Journal of Medicine* when her task was, in her words, "to take the published research articles and condense them and simplify them so that people could read them, people who were not M.D.s or even Ph.D.s." This experience translates to the specific professional goals she now holds for her students: "I learned how to write concisely in those two years. I learned how to make every word count. It was the most powerful writing experience of my life. And I think it's why that's what I drive home to these students more than anything else." The utility of this message for students' development of discursive identities of scientists is in question, however, particularly for less experienced students who strongly believe that the goal of scientific writing is mainly to clearly convey scientific findings, not to engage in a rhetorical process. The case studies that follow show that as students grapple with their shifting student identities, developing sophisticated scientific identities is a challenge.

Case Study 1: Nira—Learning to Write for a Specific Reader

A sophomore biology major at the start of this study, Nira intended to pursue a career in research and had already had several experiences working in biological research labs. Still, she felt she had few opportunities to learn to write scientifically beyond formulaic high school lab reports and that as "a technical writer, I'd say I'm pretty mediocre." In her initial survey, Nira described the communication skills needed by biologists in rhetorical terms or that biologists needed to know "how to convince an audience of the importance of their research and how to present their research in an accurate fashion." By the conclusion of the term, Nira felt she had learned a great deal, particularly about making her writing more concise, and she had a solid "foundation" for future writing, but at the same time she was not sure if what she had learned was particular to SciComm or, more specifically, to the extensive feedback she received from her SciComm instructor. Would these lessons apply to other scientific readers and other writing tasks? Had Nira started to form a discursive identity that could be applied to future scientific writing? Nira hoped so, but she was not certain.

In her start-of-term interview, Nira described successful scientific writing as that which reached a broad audience, perhaps reflective of her few experiences with such writing. "I would characterize [successful scientific writing]," she said, "as structured in that it starts off in a general sense, but specific enough to the topic so that any field can understand the idea of the project to begin with, and then as it gets more specific, it's specialized to that field." This response mirrors her more notable experience in writing for her high school science teacher, one in which the teacher's lack of familiarity with the topic meant that Nira needed to write for a more general audience and add "a lot of background in my paper, which ended up being more of, like, I do not know, not quite a scientific paper, but just like a lot of review and background." Nira was quite aware that this writing was a school-based rhetorical situation (i.e., written for her teacher) rather than mirroring the authentic task of professional scientists. At the end of the term, when she was reflecting on what she had learned in SciComm, she noted that her previous scientific writing was "more oriented on extraneous details that do not really matter too much" and that the reason for this approach was "because that's what gave me the A in high school, basically."

Doing away with these "extraneous details" seemed to be Nira's primary take-home message from her SciComm experience, and as noted previously, this lesson was the goal of her instructor. On her end-of-term survey, Nira identified "conciseness" as the most useful thing she learned in SciComm, and in her end-of-term interview, she expanded on this response, noting that she had learned to edit out "extraneous sentences or words and, like, trying to condense as much as possible and focusing on topic sentences, mainly; really like it kind of changed my whole style of writing."

In the first draft of her SciComm introduction with instructor comments in box 1.1, Nira presents a fairly long (742 words) and fairly general overview of the topic of DNA polymerases, their function and structure, and a somewhat slight idea of the specifics of her research project (though it is important to note that she wrote this draft before engaging in most of the experimental work itself). She described her process for this draft: "I didn't expect it to be a good draft. When I was reading it, I knew that it was disjointed, and it wasn't really what I wanted to say, but I didn't know how to fix it at the time, so I did the best I could, and then I knew it was a rough draft, so I handed it in and waited for the feedback."

The feedback Nira received from Ogren focused primarily on tightening her language, forecasting more clearly to the reader why she moved in particular directions within the text through the use of strong topic sentences, and tying the explanation more strongly to the actual work in the lab. Based on this feedback, Nira wrote another

Box 1.1

Nira's SciComm introduction, first and revised drafts

First draft with instructor's comments

DNA polymerases are perhaps the most important and the most well characterized *cellular* enzymes within a cell. They are responsible for the accurate and efficient replication of a cell's DNA, allowing ~~for~~ [*unnecessary*] regulated cell proliferation (Kuroita, 2005). This makes polymerases not only essential for the cell but also incredibly useful in research. [*reorder the "not only" ... "but" device—is it really needed or effective here?*]

DNA polymerases are widely used in molecular biology applications ~~today.~~ largely because of Polymerase chain reaction (PCR). [*Make your topic sentence more substantial and make it reflect the actual topic of the paragraph.*] *PCR* is a key tool for obtaining amplified target sequences in order [*see tips*] to further study these sequences in greater detail, and analyze the proteins we wish to study [*too vague*]. ~~Since~~ [*Because*] PCR is a chain reaction, ~~this indicates that~~ the replication of DNA is exponential. ~~It's therefore necessary to use a p~~*P*olymerase*s* with ~~a~~ high fidelity ~~in order to~~ [*wordy*] minimize the mutations that get perpetuated with every cycle of amplication. ~~Thus, it is important to~~ *This study focuses on* find*ing* ways to increase the fidelity and stability of such useful enzymes.

All polymerases contain three basic domains, the Thumb, the Palm, and the Fingers, and each domain is responsible for a different aspect of replication. The palm domain is conserved throughout three ~~different~~ families of polymerase, and

Final draft

DNA Polymerases are responsible for the accurate and efficient replication of a cell's DNA, allowing regulated cell proliferation (Kuroita, 2005). The enzyme's intrinsic physical properties determine its accuracy and efficiency, and its ability to replicate DNA makes it very useful for molecular biology. As a result, polymerases are perhaps the best characterized cellular enzymes known, and are essential for research.

DNA polymerases are widely used in molecular biology applications, largely because of polymerase chain reaction (PCR). Polymerases are characterized by their fidelity, which refers to the enzyme's accuracy of base pairing, and processivity, which refers to the length of template it is able to replicate before falling off. PCR is a key tool for obtaining amplified target sequences to further study the sequences and the products associated with the sequences. PCR is a chain reaction, because the replication is exponential. Polymerases with high fidelity minimize the mutations that get perpetuated with every cycle of the amplification. This study focuses on increasing the fidelity and processivity of these useful enzymes.

Polymerases are capable of polymerization and proofreading. Polymerization extends the primer 5′ to 3′ down the single template strand, creating a duplex DNA strand. Proofreading is the exonuclease activity of the enzyme, carried out 3′ to 5′, to repair mismatches that occur during polymerization. Polymerases that can exhibit both of these functions have

Box 1.1
(continued)

~~has been observed to be~~ [*wordy passive voice*] *is* [*strong verb*] responsible for catalyzing the phosphoryl transfer reaction. The structure of the thumb and finger domains vary ~~between different~~ *with* families of enzymes. The finger domain is responsible for positioning the dinuceotide triphosphates (dNTP), and the thumb is responsible for correctly position the duplex strand of DNA (Steitz, 1999).

~~There are~~ [*always reconsider—usually unnecessary*] ~~t~~*T*wo functions that can be carried out within a polymerase enzyme *are polymerization and proofreading*. ~~The first function is~~ ~~p~~*P*olymerization~~,~~ ~~extension of~~ *extends* the primer using the template strand. This occurs 5′ to 3′, down the single template strand, creating a duplex DNA strand. ~~The other function is~~ ~~p~~*P*roofreading~~, or~~ *is* the exonuclease activity of the enzyme*, and* ~~. This~~ is carried out 3′ to 5′, in order [*see tips*] to repair mistmatches that occur during polymerization. Polymerases that can exhibit both of these functions have an increased fidelity, as compared to polymerases with just replication activity. [*Make sure you define fidelity and processivity correctly.*] The two subunits, the polymerization unit and the exonuclease unit, make up what is called the Klenow fragment, and comprise the active sites of the enzyme (Steitz, 1999).

Two families of polymerases have been studied extensively [*why?*]: ~~these are~~ pol 1 and pol alpha polymerases. ~~The first family,~~ [*unnecessary*] pol 1, or DNA polymerase A, includes polymerases isolated from *Escherichia coli*, a *Bacillus* polymerase, and a bacteria called *Thermus aquaticus*. ~~The~~

an increased fidelity, as compared to polymerases with just replication activity. The two subunits, the polymerization unit and the exonuclease unit, make up what is called the Klenow fragment, and comprise the active sites of the enzyme (Steitz, 1999).

All polymerases contain three basic domains, the thumb, the palm, and the fingers, and each domain is responsible for a different aspect of replication. The palm domain is conserved throughout different families of polymerases, and is responsible for catalyzing the phosphoryl transfer reaction; it also contains the exonuclease domain. The structure of the thumb and finger domains vary with families of enzymes. The finger domain is responsible for positioning the dinucleotide triphosphates (dNTP), and the thumb is responsible for correctly positioning the duplex strand of DNA for polymerization (Steitz, 1999).

KOD DNA polymerase, a pol α archaeic polymerase, is used in laboratories because of its processivity and fidelity. It is derived from the bacteria *Thermococcus kodakarainsis* (as reviewed in Kuroita et al., 2005). Kuroita et al. (2005) discovered a mutation in the H147 residue of this protein that affected the 3′-5′ exonuclease activity of the enzyme while keeping the PCR and fidelity of a wild type polymerase. This residue lies on the tip of the thumb portion of the enzyme, as shown in Figure 5. In further studies, Hashimoto (2001, as reviewed in *Pfu* module, 2008) was able to engineer a KOD polymerase with a H147K mutation that actually

Box 1.1
(continued)

~~second family,~~ pol alpha, or DNA polymerase B, include all eukaryotic replicating DNA polymerases as well as polymerases from phage T4 and phages RB69 (Steitz, 1999). Pol alpha polymerase also include ~~what are termed~~ [*needless words*] as archaeal DNA polymerases, and have been increasingly studied for use in laboratory applications [*Why?*] (Uemori et al. 1997, Takagi et al. 1997, Braithwaite et al. 1993, Bult et al. 1996, as ~~referenced~~ *reviewed* in Kuroita et al., 2005).

Among this second class of polymerases is the polymerase from *Thermococcus kodakarainsis* (KOD DNA polymerase), highly used in laboratories because of its high efficiency and extension rate [*Can you reconstruct this sentence to make it more direct?*] (Takagi et al., 1997, as ~~referenced~~ *reviewed* in Kuroita et al., 2005). Kuroita et al. (2005) discovered a mutation in the H147 residue of this protein that affected the 3′-5′ exonuclease activity of the enzyme while keeping the PCR and fidelity of a wild type polymerase. This indicates that the mutation in this residue affected the exonuclease active site (termed the E-cleft). Hashimoto (2001, as ~~referenced~~ *reviewed* in *Pfu* module, 2008) was able to engineer a KOD polymerase with a K147K mutation that actually resulted in ~~an~~ improved exonuclease activity compared with the wild type protein. It is not known if a mutation in this site in relative [*what's this? Related?*] proteins causes a similar effect.

The KOD polymerase is closely related to Pfu polymerase, isolated from the archaeabacterium *Pyrococcus furiosus* [*Please* resulted in improved exonuclease activity compared with the wild type protein.

We believe that the H147 residue is an important site in a closely related enzyme, *Pfu* polymerase. *Pfu* was isolated from the archaeabacterium *Pyrococcus furiosus* (as reviewed in *Pfu* module). This polymerase is useful for DNA amplification in sequences up to 25 kb, with an error rate up to 10 fold lower than the bacterial-derived *Taq* polymerase (Debyser et al., 1994, as reviewed in Cline et al., 1996). This study is an effort to produce a mutation in the H147 residue that will increase the fidelity and processivity of the enzyme function. We've inserted a H147A mutation and we've tested the mutation by analyzing the processivity and fidelity of the mutant enzyme. The fidelity assay was inconclusive, while the processivity assay displayed a phenotype similar to wild type. The compilation of class data revealed possible mutants that displayed an increase in processivity.

Box 1.1
(continued)

write a better topic sentence.] (Hashimoto, 2001, as referenced in *Pfu* module). *Pfu* DNA polymerase has, specifically, been useful in high-fidelity amplification of DNA sequences up to 25 kb in size, and the error rate of *Pfu* has been found to be up to 10 fold lower than the bacterial-derived *Taq* polymerase (Debyser et al., 1994, as referenced in Cline et al., 1996). [*Break up*]

We believe that the H147 residue is an important site in the *Pfu* polymerase's structure and function [*good focus*]. ~~In order to determine this,~~ [*see tips*] ~~w~~*W*e have mutated the H147 residue into a random amino acid*, and* ~~. We've~~ tested four of these mutations by analyzing the activity of the enzyme produced by each of the plasmids above. ~~, and we are presenting our methods and our results from these assays.~~ [*Forecast our results once you get some.*]

Good start [Nira]. The content is pretty much on target. Background is ok and focus, too, but where is the justification? Please work on using fewer words to say more. This requires careful word choice. I've provided several examples. Also be sure your paragraphs focus on a single topic that is made explicit in the topic sentence.

draft, which received another round of comments (not shown), and then produced her final draft. This final draft was shorter than the first by 17 percent (613 words compared to 742), and it was much more focused on the intent of the experiment, the context for that intent, and the potential payoffs—in Swale's (1990) terms, the elements of focus, context, and justification essential to professional scientific introductions. It also offered a brief idea of the results that Nira obtained in lab. In many ways, then, Nira responded to the specific feedback she received and improved her introduction (and the rest of her paper, which underwent a similar process) as a result of that feedback.

In her end-of-semester interview, Nira looked back on her SciComm experience as a positive one, but she also identified her writing for a specific reader—and an inability to extend from that reader to other rhetorical situations—as a potential problem. In a sense, she wondered if she had developed a discursive identity as a scientist applicable beyond her SciComm instructor. As she described the writing she would do "on her own," she noted that

> while I was doing the SciCom paper, there were a lot of things I didn't like about it, particularly that I didn't know what to include and what not to include, and it seemed like even the decisions that I made, like, they were either correct or not correct, not tailored to my own choices. So I think that in the future when I write my own, it's going to be different because I'll have to decide what is important and what's not important. I will not have, you know, grades docked if I have something in there that I think is important, but they do not think is important.

In a sense, Nira expected these new rhetorical situations to be free of the trappings of school, in which her identity as a student and her grades play a key role, and thus her revision was tailored to what she felt would earn the highest grade, a prominent theme in this research. However, Nira also speculated about these new rhetorical situations and seemed to look forward to them, while knowing that she was not quite ready to face them yet:

> I think my problem at the beginning of the term was I was writing too much for a wide audience and not enough for a specialized audience. That will come with more study of the science, I think. I tried to do that as much as possible this term, but again with the requirements that we were supposed to have in our paper, I wasn't sure what was considered specialized and what was considered, like, wide. So, I'm realizing now that it's not a matter of writing style, it's more a matter of knowing the field and knowing the science.

This statement seems a particularly important moment in Nira's emerging identity as a scientific writer. As she learns the field and the science and not merely the content but the context and rhetorical demands of those contexts—as the tasks become more authentic in her view—she hopes to take forward what she learned from SciComm.

Case Study 2: Carla—Searching for a Professional Identity

Like Nira, Carla was a sophomore biology major at the start of this study and similarly imagined she would pursue graduate work in science, specifically biological engineering, following her MIT degree, though she also was considering becoming a high school biology teacher. Carla also had little experience as a scientific writer, declaring in her initial survey that what she learned from her previous scientific writing experiences was "nothing really. I felt that it was more of a teaching exercise." However, unlike Nira, Carla considered her writing background strong with a fairly developed identity as a writer, largely through her minor in and love of history, particularly ancient history, and that her writing did not have the excess verbiage that Nira felt was a problem. Instead, Carla said her writing "tends to be concise and easily understood." If anything, this concision was a problem, as Carla felt "I tend to write in a manner where the reader has to fill in the gaps."

Another similarity between Nira and Carla is that Carla also saw her SciComm experience as providing a strong "foundation" on which she would build her future scientific writing experiences. However, her commitment to biology as a career or to her future identity as a biologist seemed to play a role in her performance over the semester. Her effort and low final grade in SciComm was reflective of this uncertainty, and by the end of the term, she was ready to move on to different challenges.

In terms of Carla's identity and future as a science student, her circuitous route to biology at the start of her sophomore year was perhaps indicative of the fleeting nature of that decision:

> What I really wanted to when I came here was [biological engineering], and then I took [a biological engineering class] last semester, and I didn't really actually just have a good time in that class. I felt like the department was very new. Like I wasn't really standing on solid ground with a lot of the TA's, a lot of the professors, and then I knew I wanted to do grad school anyway, so then I just made a decision to switch from [biological engineering] to [biology]. On top of that, I took a [humanities] class last semester, which really made me want to major in history as well, so now I can double major in biology and history and still do grad school for biological engineering.

She added in her start-of-term interview that biology had always been "the subject I'm just best at, too," which factored heavily in her decision. A powerful factor in Carla's learning in SciComm, then, was her shifting identity as a student and her search for a disciplinary comfort zone—whether that comfort would come from academic success or congruence with her future plans.

In terms of Carla's conception of successful scientific writing at the start of the term, she offered a rhetorical situation in which conciseness and clarity were in the service of reaching a fairly general audience: "You always have to, like, state everything you did, state why you did it and, then, like, be clear and concise so anyone from any field can understand what you're saying." Carla followed this belief closely when she revised the draft of her SciComm introduction, reporting that she solicited opinions from one friend who had some knowledge of biology and from another who was unfamiliar with the field to make sure that "the intro was as perfect as possible."

The issue with the writing task for SciComm, however, was that it was quite particular to the discourse community of Experimental Biology, given the specific content about DNA polymerase and the knowledge of the research in the field that one needed to know, as well as the particular demands of her SciComm instructor. In her end-of-term interview, Carla expressed some dismay with the practice in SciComm to critique poor models of scientific introductions and not to offer many ideal models: "The intro I felt was very, very difficult just because what I usually base my writing on . . . professors will usually give, like, this is a good example of what somebody did, and I felt like the examples that we had were just poor examples, especially because we discussed in class how they were poor examples, and I just found it very difficult." In other words, Carla's quest for "perfection" in her introduction was frustrated by her seeking out preliminary readers who could not represent what Ogren would value as a scientific professional and a writing teacher and by Carla's lack of experience with this kind of scientific writing task.

Carla's first draft for the introduction in box 1.2 can be characterized as in accord with a feature she described in her writing: it is lacking in the specific detail required of this task. Nevertheless, her final version does not show much change, only the incorporation of most of her instructor's recommended edits and the addition of one sentence of background on thermostable polyermases. Such minimal reworking in this and subsequent components resulted in Carla's receiving one of the lowest grades among students in her SciComm section .

Also like Nira, Carla seemed to feel that she was writing to one specific reader, her SciComm instructor, rather than to a more general scientific audience and that her lack of experience with this task resulted in a great deal of uneasiness. She summarized her experience as follows: "SciComm was a little excruciating because I had never written like that before, so I was just, like, I do not really know what I'm doing, especially the discussion section. The discussion section I felt like I turned in one and I was like, yeah, and I got like a check minus on it and I'm like, what?"

Despite her grade, by the end of the semester, Carla reported a great deal more comfort with the genre of the research article and that concise, hypothesis-driven writing had filtered over into the work she was doing in a political science course. But the end of the semester also brought a change for Carla: she had decided to change her major

Box 1.2
Carla's SciComm introduction, first and revised drafts

Introduction First draft with instructor's comments (added words in italics; comments in bracketed italics)

DNA polymerization is part of [*needs article*] mechanism by which DNA replicates itself [*needed?*]. Specifically, it is the act of adding additional nucleotides to the replicating string of DNA by the assistance of enzymes known as DNA polymerases. A DNA polymerase's general exonuclease activity can be described by its processivity, ~~or~~ the speed of assembly and fidelity, ~~or~~ the accuracy of assembly. ~~Currently, the scientific community seeks to~~ [*empty words—get right to the point*] ~~i~~*I*ncrease~~e~~*ing* the processivity and fidelity of the commonly used DNA polymerases ~~because~~ *would benefit* many standard laboratory procedures, such as Polymerase Chain Reaction (PCR)*, which* demand fast and accurate DNA replication (Kuroita et al., 2005).

~~In a recent study~~, [*Avoid long lead—bogs down the reading—get right to the point*] Kuroita et al. (2005) *recently mutated* the DNA polymerase Thermococcus kodakaraensis [*ital*] (KOD1) ~~was mutated~~ by substituting the 147 amino acid position Histidine with a Lysine. This mutation resulted in an increase in ~~exonuclease activity by higher~~ processivity and fidelity than the wild type. ~~This~~ *The present* study focuses on mutating the *Pyrococcus furiousus* (*Pfu*) DNA polymerase to increase its processivity and fidelity ~~from wild type in a manner similar to that has already been accomplished with KOD1 DNA polymerase~~ [*adds no new info*].

Introduction final draft

DNA polymerization is part of the mechanism by which DNA replicates. Specifically, DNA polymerization is the act of adding additional nucleotides to the non-template strand of DNA. This is done with the assistance of enzymes known as DNA polymerases. DNA polymerases are defined by two qualities: processivity—the speed of assembly, and fidelity—the accuracy of assembly. Increasing the processivity and fidelity of the commonly used DNA polymerases would benefit many standard laboratory procedures, such as Polymerase Chain Reaction (PCR), which demands fast and accurate DNA replication (Kuroita et al., 2005). Typical PCR requires a thermostable DNA polymerase, a DNA polymerase that can function at 100 degrees Celsius for several minutes (7.02 Lab Manual, 2008).

Kuroita et al. (2005) mutated the thermostable DNA polymerase, Thermococcus Kodakaraensis (KOD1) by substituting the 147 amino acid position Histidine with a Lysine. This mutation resulted in an increase in processivity and fidelity compared to the wild type. This study focuses on mutating the thermostable *Pyrococcus furiosus (Pfu)* DNA polymerase in an effort to improve its processivity and fidelity.

The DNA polymerase *Pfu* is strikingly similar in structure to KOD1. The two DNA polymerases are approximately 90% identical in amino acid sequence. KOD1 and *Pfu* also have similar half lives and proofreading capability. However, they differ in extension rate and accuracy. *Pfu*

Box 1.2
(continued)

The DNA polymerase *Pfu* is strikingly similar in structure to KOD1. The two archael DNA polymerases are approximately 90% identical when comparing amino acid sequences. ~~In many respects~~ [*needless words*] KOD1 and *Pfu also* have similar properties. Both DNA polymerases for example have the same half life and proofreading capabilities. Where they differ is in extension rate and accuracy. *Pfu* has an extension rate 1/3 the rate of KOD1, but has a accuracy 1.75 times that of KOD1. ~~This study seeks~~ *We hypothesize* [...] to keep the high accuracy of wild type *Pfu* polymerase while increasing its exonuclease activity. [*Please briefly describe your approach (a sentence or two). When you get your results, add a line or two to forecast them.*]

Good start [Carla]—Please focus on word choice and sentence structure to make your writing more compact. Also be aware of needless phrases and sentences. You could include some background on thermostable polyermases. Use your lab manual as a source of background as well—you can cite it.

has an extension rate 1/3 the rate of KOD1, but has a accuracy 1.75 times that of KOD1.

We aim to keep the high accuracy of wild type *Pfu* polymerase, while increasing exonuclease activity. Using site-directed mutagenesis, we substituted the *Pfu* 147 position Histidine with a Tryptophan. Our study is consistent with the published literature in that the H147W mutant has a greater processivity than wild type.

to materials engineering with biology as her minor. As she described this decision, "I've pretty much taken what I've wanted, I guess, out of biology, and I'm like, I do not know, I'm very fickle so, like, I just was like, all right, I got biology. I got a sense of it, saw what it's about. I want to do something new now."

Although Carla did see the SciComm writing tasks as ultimately useful, the shifting nature of her future plans and of her identity as a science and engineering student makes the lasting impact relatively unresolved. Carla's case also raises questions about developing a discursive identity when one's future professional identity itself in is flux. How can students' scientific discursive identities be flexible enough to serve them in a variety of classes in a variety of disciplines? How much is learning indexed to specific tasks and contexts? In SciComm, the writing task was relatively specific to molecular biology and to the science of DNA polymerase. For students such as Carla with only a weak commitment to that field and uncertainty over her future, distilling larger lessons from this specific task was difficult.

Case Study 3: April—Facing New Learning Challenges

Carla was not the only SciComm participant with a future in relative flux. April, a sophomore chemical engineering major, was only starting to develop a sense of her future paths and of scientific writing as a rhetorical act. When asked about her plans for her career after MIT, she speculated about possibilities in the pharmaceutical industry or in biotech, but largely confined her vision to a summer internship possibility. When asked about her conception of what scientific writing entails, her response focused on the notion of experimental results supported by evidence: "I do not know if you need to be, like, highly persuasive, like it's not a debate, but I think just having to convince your readers that you actually know what you're talking [about], I guess providing enough backup in your analysis or something, just so you're not, like, you're not just pulling your analysis out of anywhere. You are saying why you think that result happened, that kind of thing."

April's uncertainty with a scientific discursive identity extended to the writing she might encounter as a chemical engineer or the types of skills she might need. When asked to imagine that future, her answer indicated a fairly limited idea of those possibilities, marked, as she acknowledged, by her lack of direct experience: "It's a good skill to have, in general, to be able to perform an experiment and get the results and be able to tell it to someone else. I guess that's definitely something good to know how to do. I'm sure I'll have to write papers. I do not know how many. I'm sure it would probably be less than, like, a bio major would have to or something, but I do not really know that yet."

In terms of her introduction to her SciComm paper, April's first draft with instructor's comments in box 1.3 is not appreciably different from Carla's with the exception of her confusion over amino acids, rather than nucleotides, as the basic elements of DNA. In her rewrite, she incorporates the suggestions of her SciComm instructor without going beyond those suggestions to tighten her wording and add a bit more background literature. Overall, it is a solid middle-of-the-pack effort, one that led April to offer in her final interview that "I feel pretty confident about being able to write a good paper now" (see box 1.3).

While April might have felt "pretty confident" as a scientific writer at the end of term, her discursive identity or her conception of what she learned from SciComm

Box 1.3

April's SciComm introduction, first and revised drafts

Introduction first draft with instructor's comments

All life forms store genetic information-including information on behavior- in DNA. DNA is formed by double-stranded chains of amino acids, and a small change, or mutation, in the sequence of those amino acids can create profound changes in the behavior of an organism. Each cell needs DNA to direct its growth and function. DNA replication, which occurs just before cell division, ~~is necessary so that~~ [*wordy*] *provides* each new cell ~~can have~~ *with* an identical copy of the parent DNA strand. This replication is performed by enzymes called DNA polymerases. During replication, the DNA splits into single strands, and the DNA polymerase "reads" these strands and adds amino acids to the 3′ end of the complementary strand forming along the parental DNA strand to create identical copies of the parental strand. Different types of DNA polymerase ~~exist from different organisms, and a polymerase is defined by its~~ [*wordy*] *are characterized by their* processivity (speed at which it adds new nucleotides to the new strand) and fidelity (accuracy of ~~the process~~ *replication*, and error correction). [*good context*]

~~To perform~~ ~~e~~Experiments using DNA~~,~~ ~~we must have~~ *require* multiple copies of the DNA available. [*Try writing direct sentences that are right to the point.*] ~~We can use the~~ [*needless* words] ~~p~~Polymerase chain reaction [*strong subject*] ~~, or~~ (PCR)~~,~~ ~~process to~~ creates [*strong verb*] multiple copies of DNA ~~to use for study~~. This

Introduction final draft

All life forms store genetic information—including information on behavior—in DNA. DNA is formed by double-stranded chains of nucleotides, and a small change, or mutation, in the sequence of those nucleotides can create profound changes in the behavior of an organism. DNA replication, which occurs just before cell division, provides each new cell with an identical copy of the parent DNA strand or gene. This replication is requires enzymes called DNA polymerases. During replication, the DNA splits into single strands, and the DNA polymerase "reads" these strands and adds nucleotides to the 3′ end of the complementary strand forming along the parental DNA strand to create identical copies of the parental strand. Different types of DNA polymerase are defined by processivity (speed at which it adds new nucleotides) and fidelity (accuracy of replication, and error correction).

Experiments using DNA require multiple copies of that DNA. Polymerase chain reaction (PCR) rapidly creates multiple copies of DNA. This process occurs at high temperatures, requiring a polymerase that does not denature at high temperatures. The DNA polymerase of Thermococcus kodakarensis (KOD), an archaeal strain of bacteria, can operate at the high temperatures required for PCR. Researchers discovered that a single KOD mutation can cause dramatic improvements in the processivity and fidelity of this polymerase. When histidine, amino acid 147, was substituted with other amino acids, the

Box 1.3
(continued)

process occurs at high temperatures, and ~~we need~~ *requires* a polymerase that ~~will~~ *does* not denature at high temperatures. ~~One such polymerase has been found, in~~ [*needless words*] Thermococcus kodakarensis (KOD) *is a . . . [describe].* Researchers discovered that a single *KOD* mutation *can* caus~~ed~~ dramatic improvements in the processivity and fidelity of ~~the~~ *this* polymerase [*This sentence could be more specific*]. When histidine, amino acid 147, was substituted with other amino acids, the effects were varied (Kuroita et al., 2005).

We mutated a similar polymerase~~,~~ from the archeabacterium *Pyrococcus furiosus* (*Pfu*)~~,~~ to determine the effects of the mutation on ~~the~~ *this* enzyme. ~~Our~~ *We* hypothes~~is~~*zed* ~~is~~ that mutation of the H147 gene ~~will~~ *could* create a change in the processivity ~~and~~ *or* fidelity of the *Pfu* DNA polymerase. We ~~want to study if those~~ [*This is process-oriented langauge—avoid.*] *examined whether such* changes increase or decrease the function of the enzyme.

Good start, [April]. The content is pretty much on target but your sentences tend to contain several needless words & phrases as indicated. You might include the study by Hashimoto (KOD) in your background info.

effects were varied (Kuroita et al., 2005). Lysine substitution caused the greatest improvements in polymerase and exonuclease activity, and a lower mutation frequency.

We mutated a similar polymerase, from the archeabacterium *Pyrococcus furiosus* (*Pfu*), in an effort to improve the function the enzyme. We hypothesized that mutation of the H147 gene could create a change in the processivity or fidelity of the *Pfu* DNA polymerase. We examined whether such changes increase or decrease the function of the enzyme.

was focused on the rudiments of structure and format. In her end-of-term survey, April identified the most useful thing she learned in SciComm as "how to format a typical scientific paper" and that while she was writing her SciComm paper, she felt that she was learning "more of the format of a research article than anything else." April's relatively narrow conception of what one learns by writing about science also meant relatively low performance in SciComm and in Experimental Biology as a whole. Her final SciComm grade was fifteen points below class average, and her final Experimental Biology grade was close to the bottom of the class.

Although April ultimately felt that the task of the SciComm paper was useful, her lack of clear vision about a professional future complicates the relationship between the task itself and her emerging identity as a scientist. As her overall grade in Experimental Biology showed, April was struggling with the science itself. Ogren identified this uncertainty and relative timidity in her end-of-term reflections: "[April] seemed to struggle with the content, and she struggled with the writing. She was just desperate to do what was right. . . . She was just desperate to figure out what she was supposed to do and get it, get it right." April presents a challenge to notions of students developing discursive identities when they are struggling to grasp the content of the science they are writing about. Is there a developmental threshold for students in order for them to derive maximum benefit from authentic tasks such as research articles? Or are the benefits not necessarily apparent by the end of a single semester? April's case raises these questions and complications in the relationship between writing science and developing discursive identity.

Case Study 4: Jake—Stepping Stones to Success

Jake, the lone senior among the SciComm research participants, had the most extensive experience with scientific research and with scientific writing, and, perhaps as a result, the most sophisticated understanding of what it means to write as a professional. Overall, Jake's academic accomplishments were impressive. At the start of the semester under study, he was a senior physics major with an original intention of pursuing a Ph.D. in high-energy physics. He was enrolled in Experimental Biology because during the fall semester of his junior year, he had decided to shift his future identity from "physicist to physician," in his words, and complete the course work needed for premedical training. He completed this course work in three semesters, a remarkably short period of time, took the MCAT, and scored high. By the end of his senior year, he learned he had been accepted to a prestigious MD/PhD program. In a sense, then, Jake's identity as a physician scientist was a powerful motivating device for his success in pursuit of this goal, but he also learned that he could succeed in a new field. "What

I've learned," he told us, "is that, from this shift [in career goals], is that personally and perhaps in general, ... I feel I'm able to go from one area to another, and I can learn a very different field if I have the motive, the desire."

In terms of his conception of what it meant to write like a scientist, Jake took a view in accord with his desire to do good in the world:

> There's a logical process involved in the scientific writing. And ... especially to be able to communicate. I see that as the goal of the writing. It's not just to capture your thoughts, like a brain dump ..., but it is to communicate the message that you have with an intention of educating those that you're writing to in your audience—and hopefully inspiring them in some way so that they can be the better for it.

For Jake, a key concept in scientific writing was the difference between presentation and communication—in a sense, the difference between understanding scientific writing as mostly a matter of correct formatting and concise style versus scientific writing as a form of persuasion. In Jake's words, the purpose of scientific writing is "being able to articulate in a coherent way, intelligible way, both to the general public and to your colleagues. I mean, that's the whole point. We're not just trying to find something out just for ourselves. If we're trying to find some sort of truth or if we're trying to make a discovery, what real good is that unless it's shared and that message is communicated, not just presented?"

Jake also shared a great deal in his interviews about the differences that he saw between the extensive writing he had done as a physics students and the new kind of writing he was encountering in SciComm. By the end of the semester, some of what he had learned about scientific writing in SciComm was in contrast to the important skills he had taken away from the writing-intensive experiences of Physics Junior Year Lab: "I'm taking away how to present the scientific method in written form. That's definitely what I'm taking away, more than my Junior Lab papers. I feel that through SciComm, I've been able to develop that structure, that hypothesis driven, data-driven type analysis then relating it back to my hypothesis. I see more of the vision of how to structure these papers."

In terms of other outcomes he felt he derived from SciComm, Jake described a deeper understanding beyond mere formatting: "I came to learn that it can be more important to focus on the principles of your methods and the most important principles of your design, rather than creating a lab report that this is exactly the order in which I did everything." He also felt this learning, as opposed to what the other participants reported, was not confined to writing in biology or, more specifically, to what was required for the SciComm paper. Instead, Jake saw the larger lessons he learned in SciComm as applicable to a wide range of writing: "I think that in my papers I've greatly improved on the logic, and the organization. This whole semester, not just in Sci-Comm but in all of my writing, in my literature writing, in my history writing, but I think a large part of that has come out of the thinking in SciComm."

In terms of his SciComm research paper introduction, Jake offered a comprehensive reading of the field, the importance of the laboratory techniques using the normal activity of DNA polymerases, and the potential payoffs of the line of research he was pursuing in Experimental Biology. As he reconstructed in an interview his thinking process for his introduction, he showed his comfort with the work and with taking on the identity and using the language of a molecular biologist in pursuing this line of research: "I tied it all together into this continual search for improved DNA polymerases. Well, we're then going to use site-directed mutagenesis to study the DNA polymerases that are needed for PCR for better improved study of genetic, I guess to tie it back to my motivation in the introduction of how we're trying to mutate and explore protein function and structures."

Jake also brought in far more literature to his introduction than other students, and in Ogren's comments on his draft, she wonders if he has read these sources or had found them cited in secondary sources, which almost all other students tended to do. In his start-of-term interview and in response to her comments, Jake noted that "I always go to the source. And I do not take it for granted.... I do not like citing secondary sources. I'll go and find in the literature the original papers and quote from those and reference them." Overall, Jake scored highly on this preliminary draft, needed to make few changes to his final draft, and ended with one of the top SciComm grades in his section (see box 1.4).

Jake's success in SciComm and Experimental Biology and his academic achievements offer a contrast to the other research participants in his sophisticated view of the processes of and stakes for the communication of scientists. As Jake described the primacy of the research article format, he showed how he viewed the writing of science as a social act rather than merely a faithful rendering of the natural world. He understood format as a "mold" of sorts, then commented: "There definitely has to be a mold when you're writing some of these papers because that's how ... it's like a guild system. Journals, you go into the guild system, and you have to do it their way for you to get to the top of the guild. But once you're the master of the guild, then you can define the practice."

It would not be a stretch to imagine Jake as "master of the guild" someday as an editor of a scientific journal and a key shaper of what scientific writing might look like or what constitutes authentic texts. But those activities would likely be in the interest of the advancement of science and its impact in the wider world. Jake described

Box 1.4
Jake's SciComm introduction, first and revised drafts

Introduction first draft with instructor's comments

The genomes of increasingly sophisticated organisms, including humans, have been sequenced, leading to examination of implications of genomics on proteomic expression, cellular development, and the inheritance and *patho*physiology of human diseases. Genes have been characterized by expression in isolation, and their protein function elucidated by various techniques, including mutation. This characterization and cataloguing of genetic information~~,~~ *and* protein structures ~~and~~ across species is allowing molecular biologists and geneticists to probe patterns of conservation in evolution [*among other things*], ~~understand~~ recombinant DNA methods [*seems misplaced*]. In addition, developments such as recombinant DNA techniques and gene 'knock-out' during homologous recombination and other methods permit ~~engineering of~~ protein expression *engineering* and proffer future medical treatments, such as gene therapy [*The methods are important for showing targets for drug therapy.*]. ~~Simultaneously,~~ ~~t~~*T*hese developments have required the ability to create large quantities of synthetic DNA from samples as small as single molecules, with efficiency, low cost, and high fidelity.

A primary method of *in vitro* DNA synthesis, the polymerase chain reaction (PCR). PCR controls the duplicative machinery of Nature [*?*], DNA polymerases, to replicate DNA. The reaction first melts double-stranded DNA to single strands

Introduction final draft

The genomes of increasingly sophisticated organisms—including humans—have been sequenced, leading to examination of implications of genomics on proteomic expression, cellular development, and the inheritance and pathophysiology of human diseases. Genes have been characterized by expression in isolation, and their protein function elucidated by various techniques (including mutation). This cataloguing of genetic information and protein structures across species is allowing molecular biologists and geneticists to probe patterns in evolutionary conservation, match homologous genes with function, and associate mutations with disorders. In addition, developments such as recombinant DNA techniques and gene 'knock-out' during homologous recombination may enable both protein expression engineering to identify genetic targets of disease and gene therapy to correct them. These developments have required the artificial synthesis of large quantities of DNA from samples as small as single molecules, with efficiency, low cost, and high fidelity.

A primary method of *in vitro* DNA synthesis, the polymerase chain reaction (PCR), manipulates nature's duplicative machinery (DNA polymerases) to replicate DNA. The reaction first melts double-stranded DNA to single strands (95 deg C), anneals 3′-5′ reverse and 5′-3′ forward primers to the ss DNA (56 deg C) by complementary binding, then extends the primers as *in vivo* by the 5′-3′ endonu-

Box 1.4
(continued)

(95 deg C), anneals 3′-5′ reverse and 5′-3′ forward primers to the ss DNA (56 deg C) by complementary binding, then extends the primers as *in vivo* by the 5′-3′ endonuclease activity of DNA polymerases in a solution of dNTPs (at 72 deg C) (Saiki, R., Mullis, K., et al., 1988 [*Did you read the original article?*]; see also Nobel Lecture of Kary Mullis, 1993). A single cycle duplicates the DNA, while n cycles amplify~~ies~~ the native DNA sample (X0) exponentially, by $X(n) = X02^{n}$. The 95 degress Celsius condition to thermodynamically dissociate DNA base-pair hydrogen bonds denatures normal proteins, and necessitates either regular input of DNA polymerases or thermostable alternatives. *Hyperthermophiles* of the kingdom Archaea, such as *Thermus aquaticus* found in deep water hydrothermal vents or geysers, have thermostable polymerases (e.g., [*e.g., requires a comma*] *Taq*) with optimal function at up to 80 degrees Celsius. The continued selection and development of such highly-processive and thermostable polymerases for commerical PCR is of great interest (Cline et al., 1996). In addition, identification of small genomic variations, such as mutations characterizing disease or single-nucleotide polymorphisms (SNPs) associated with parasite drug resistance, requires high fidelity DNA amplification.

One particular technique expanding our understanding of genomics is site-directed mutagenesis (Smith, 1985 *[Did you read the original article?*]; see also Nobel Lecture in Chemistry 1993). This process manipulates the PCR amplification

clease activity of DNA polymerases in a solution of dNTPs (at 72 deg C) (Saiki, R., Mullis, K., et al., 1988; see also Nobel Lecture of Kary Mullis, 1993). A single cycle duplicates the DNA, while n cycles amplify the native DNA sample (X0) exponentially, by $X(n) = X02^{n}$. The high temperature (95 deg C) needed to thermodynamically dissociate DNA base-pair hydrogen bonds denatures normal proteins, and necessitates either regular input of DNA polymerases or thermostable alternatives. *Hyperthermophiles* of the kingdom Archaea, such as *Thermus aquaticus* found in deep water hydrothermal vents or geysers, have thermostable polymerases (e.g., *Taq*) with optimal function at up to 80 degrees Celsius. The continued selection and development of such highly-processive and thermostable polymerases for commerical PCR is an imperative for efficiency in biotechnology (Cline et al., 1996). In addition, identification of small genomic variations, such as mutations characterizing disease or single-nucleotide polymorphisms (SNPs) associated with parasite drug resistance, requires high fidelity DNA amplification.

One particular technique expanding our understanding of genomics is site-directed mutagenesis (Smith, 1985; see also Nobel Lecture in Chemistry 1993). This process manipulates the PCR amplification of DNA by use of synthetic forward and reverse primers with customized central mutations surrounded by site-complementary pairs. The mutated primer anneals to the template ssDNA of interest, is elongated by polymerases, then subse-

Box 1.4
(continued)

of DNA by use of synthetic forward and reverse primers with customized central mutations surrounded by site-complementary pairs. The mutated primer anneals to the template ssDNA of interest, is elongated by polymerases, then subsequent cycles linearly amplify the mutant DNA; the mutant protein of interest can then be studied following cloning and expression (e.g.) by microorganisms. This permits study of the direct relationship of genetic sequence to protein structure to protein function.

Specifically, structure characterization by X-ray crystallography or NMR spectroscopy combined with study of site-directed mutants has allowed elucidated and improvement of 5′-3′ endonuclease and 3′-5′ exonuclease mechanisms for PCR polymerases. Hashimoto et al. (2001) crystallized the family B DNA polymerase of the archaeon *Thermococcus kodakaraensis* KOD1, identifying an exonuclease active cleft (E-cleft), a Palm domain, and two Thumb (endonuclease) sub-domains. It was in the unique loop of the E-cleft that Kuroita et al. (2005) found a mutant (H147K) in KOD1 that resulted in a 2.8 fold increase in 3′-5′ exonuclease activity over the wild-type enzyme. This modification of a key residue by site-directed mutagenesis improved fidelity from a mutation frequency of 0.47% to 0.12%, as opposed to 7.9% in Taq and 1.3% in *Pfu*, while maintaining superior elongation rates (130 bp/s compared to 20 bp/s for *Pfu*). Specifically, Kuroita's success emphasizes the criticality of this residue in the catalytic exonuclease mechanism. Generquent cycles linearly amplify the mutant DNA; the mutant protein of interest can then be studied following cloning and expression by microorganisms. The biologist becomes an experimentalist with active control of the genetic sequence-protein structure-protein function relationship.

Recently, detailed structural characterization (by X-ray crystallography or NMR spectroscopy) combined with study of site-directed mutants has elucidated and improved PCR polymerase 5′-3′ endonuclease and 3′-5′ exonuclease mechanisms. Hashimoto et al. (2001) crystallized the family B DNA polymerase of the archaeon *Thermococcus kodakaraensis* KOD1, identifying an exonuclease active cleft (E-cleft), a Palm domain, and two Thumb (endonuclease) sub-domains. It was in the unique loop of the E-cleft that Kuroita et al. (2005) found a mutant (H147K) in KOD1 that resulted in a 2.8 fold increase in 3′-5′ exonuclease activity over the wild-type enzyme. This modification of a key residue by site-directed mutagenesis improved fidelity from a mutation frequency of 0.47% to 0.12%, as opposed to 7.9% in Taq and 1.3% in *Pfu*, while maintaining superior elongation rates (130 bp/s compared to 20 bp/s for *Pfu*). Specifically, Kuroita's success emphasizes the criticality of this residue in the exonuclease mechanism. This work demonstrates the promise of site-directed mutagenesis to engineer native protein (or enzyme) forms for improved performance in biotechnological and industrial applications, such as high-fidelity PCR.

Box 1.4
(continued)

ally, it demonstrates the promise of site-directed mutagenesis to engineer native protein (or enzyme) forms for improved performance in biotechnological and industrial applications, such as high-fidelity PCR. *Pfu*, a DNA polymerase from the archaeon *Pyroccocus furiosis* widely used in the PCR amplification of DNA samples, is a homolog of KOD with an E-cleft domain of hypothesized homologous exonuclease function. Like KOD, modification of histidine 147 in the unique loop of *Pfu* is predicted to alter exonuclease activity, revealing similarities with the KOD exonuclease mechanism, and has similar potential for mutagenic polymerase improvements. We present data assessing the role of H147 in the 3′-5′ exonuclease and 5′-3′ endonuclease activity and mechanisms of *Pfu*, as determined by comparative assays of wild-type *Pfu* with mutant *Pfu* proteins prepared by site-directed mutagenesis of the AA147 residue. Moreover, we discuss the relevance of homology comparisons for prediction of the functional outcomes of mutants of similar proteins. *Great job, [Jake]. See comments within.*	*Pfu*, a DNA polymerase from the archaeon *Pyroccocus furiosis* widely used in the PCR amplification of DNA samples, is a homolog of KOD with an E-cleft domain of hypothesized homologous exonuclease function. Like KOD, modification of histidine 147 in the unique loop of *Pfu* is predicted to alter exonuclease activity, revealing similarities with the KOD exonuclease mechanism, and has similar potential for mutagenic polymerase improvements. We present data assessing the role of H147 in the 3′-5′ exonuclease and 5′-3′ endonuclease activity and mechanisms of *Pfu*, as determined by fidelity and processivity assays of wild-type *Pfu* compared with a cohort of mutant *Pfu* proteins prepared by site-directed mutagenesis of the AA147 residue. Moreover, we discuss the relevance of homology comparisons for prediction of the functional outcomes of mutants of similar proteins.

the responsibilities of the scientist in relation to his or her writing: "That responsibility is not just simply to communicate your thoughts but you're also responsible to whoever reads it."

It is important to recall that Jake was the lone senior among study participants. His sophistication and ease with his identity as a scientific writer are perhaps the result of his advanced class standing in comparison to the other case-study students. Thus, all four cases raise questions about the relationship between overall sociocognitive development and writing success, a factor that comes into play in the next chapter as students further along their undergraduate careers and more committed to the discipline with which they identify seem to have much more success as scientific writers.

Summary of Introduction to Experimental Biology and Communication

The goal of SciComm—to develop students' discursive identities as scientists, including knowledge of the scientific article's components, the rhetorical role of those components, and the processes by which a scientist produces an article—was achieved in varying measures for the case study participants. All four students reported high levels of satisfaction with their SciComm experiences, and all believed they had created a foundation on which future writing could stand. Implications of these results include the role of the relationship between students' views of knowledge creation in scientific writing, the ways that shifting student and career identities affect a developing scientific identity, and the strong role that school as a context (students' identities *as* students) played in their learning:

Students' view of scientific writing—whether as knowledge transfer or as rhetorical act—played a strong role in their success as scientific writers and in the class itself. SciComm students who saw scientific writing as mostly a matter of information transfer (albeit in concise and highly structured forms) tended to struggle more with the relatively authentic classroom tasks they faced. Students who could imagine the audience's needs for their writing and connect that writing to past texts and future texts tended to have more success. Jake, the lone senior, had success at taking on new roles and shifting his career goals, and this shift did not present uncertainty for him as a scientific writer. Instead, his confidence that he could take on new identities and learn new rhetorical situations (and, perhaps most important, recognize them as rhetorical situations) resulted in a strong performance. The other three case study participants had a view of scientific communication as largely information transfer and were unsure of the ways that the context of SciComm might extend to other contexts. The developmental continuum that this result indicates speaks to the need to assess students' views of knowledge production in science and to be explicit about the applicability of skills learned in a particular context to additional contexts.

Students' shifting identities and uncertain futures played a strong role in their success in SciComm. Carla's uncertainty about her major and April's lack of definite vision for her future raised questions about students' developmental readiness to benefit from authentic tasks. The need to develop rhetorical flexibility and apply lessons learned in SciComm to future scientific writing was in competition with their larger concerns about majors and careers. Jake's clear vision of his future—and the kinds of writing, speaking, and thinking that he would need to do—allowed him to optimize his SciComm experience. Certainly students who are learning to write in college often have shifting identities as students and for their postcollege careers. In fact, the writing tasks themselves can play a role in helping students develop these identities, particularly their discursive identities as scientists and engineers.

The context of schooling—students' time available, dedication to writing and revising, the presence of a grade or/evaluator, and the realization that these tasks were not quite "real"—played a strong role. For several participants, career decisions based on previous academic success and writing behaviors based on high grades in previous classes did not necessarily serve them well in SciComm. The ongoing dilemma is to separate the contribution of writing tasks to students' development of discursive identity from the assessment of those tasks as more than one teacher's individual values. However, schooling is also a laboratory for students—a place to try out new discursive roles and to receive instruction in how to write, speak, think, and act. For most students, SciComm is a preliminary step in that development, one that they will build on by subsequent course work and communication tasks (as shown in subsequent chapters of this book). Students' identities are in flux, in other words, and this state is the norm, presenting opportunities for growth and development.

2 Taking On the Identity of a Professional Researcher

Learning to acquire a discursive identity as a scientist or engineer is largely dependent on engaging in the communication practices that professionals use. In chapter 1, that practice was the scientific research article in its codified IMRD format, and the students represented a range of majors, developmental levels, and possible futures. In a sense, the in-flux nature of students' identities as neophyte scientists and the dominance of their identities as students strongly shaped their learning experiences.

In chapter 2, identity is also a key factor, but in this class, all of the students were biological engineering majors, and the tasks were more varied, as befitting the relatively new discipline of biological engineering and the wide variety of roles students might play as future leaders in it. In other words, what was important to the class profiled in this chapter is that students engaged in authentic communication tasks closely mapped to the identity of professional biological engineers. This chapter addresses two questions:

- What is the relationship between authentic writing tasks and students' development of scientific discursive identities?
- What are students' challenges and opportunities as they face the multiple writing tasks of biological engineers?

Authenticity has a long history in teaching students to communicate as scientists. Science educators have long identified the need to have students study the natural world not by reading about it in textbooks or by listening to lectures but by engaging in hands-on, authentic experiences. The challenge was described by Harvard professor of zoology Louis Agassiz in the mid-nineteenth century: "The pupil studies nature in the schoolroom, and when he goes out of doors he can not find her" (quoted in Campbell 1893, p. 119). Since that time, generations of science educators at all instructional levels have used laboratory experiments as a way to offer students the authentic experiences of scientists. The idea here is that the "aha!" moment that has propelled scientific discovery and technological progress will recur in the more controlled conditions of the school science laboratory. As a result, it is hoped that new generations of scientists will be created.

Despite this intent, laboratory advocates have long faced criticism—from the amount of resources needed for such intensive teaching (Grier 1935), to the reduction of laboratory learning to rote cookbook exercises (Hodson 1998), to the mismatch between teachers' goals and students' perceptions of their learning (Russell and Weaver 2008). Nevertheless, the larger goal of authentic learning has held strong. In a sense, the belief is that giving students the experiences of real science as practiced by real scientists will, for some students at least, lead them to scientific careers, and the imperfect match between this intent and the realities of implementation often accounts for the constant dissatisfaction with laboratory learning.

When it comes to student writing in school science laboratories, the hue and cry for authenticity is a bit more muted and a bit more complicated. Research in the rhetoric of science (Bazerman 1988) or science as a human activity (Latour and Woolgar 1986) has shown how circuitous the route is from initial idea to published result and how that route exists within myriad social forces. However, giving students opportunities to experience the processes of discovery and problem solving and then to write about those processes as would professional scientists is necessarily constrained by time, resources, entrenched teaching practices, and students' prior knowledge, among other factors (see Bazerman 1988, Lerner 2007, Zerbe 2007). Thus, while the *New York Times* could declare that the solution to the problem of inadequate science education is that "students need early, engaging experiences in the lab—and much more mentoring than most receive now—to maintain their interest and inspire them to take up careers in the sciences" ("How to educate" 2006, p. A14), writing about those lab experiences is advocated with much less certainty. When students do write about lab work, Hodson describes the all-too-common result as follows:

> Experiments are regarded as decisive tests of validity of hypotheses and conjectures. This myth is reinforced by insistence on a third person, passive voice, emotionally neutral style of laboratory report in which the experimentally determined factual evidence supposedly "speaks for itself," and any suggestion that the experimenter/inquirer is engaged in the active construction of meaning is carefully excised. (1998, p. 94)

Student writing about school-based laboratory work is still promising, however, from the point of view of apprenticeship as a type of learning. Carter, Ferzli, and Wiebe ascribe the power of the student lab report to its potential as an apprenticeship genre because "apprenticeship genres such as the lab report could play a critical role in the socialization of undergraduates into disciplinary communities" (2007, p. 295). At the same time, these authors acknowledge the apprenticeship nature of this activity:

Certainly the purpose and audience of the scientific journal article differ in important ways from the purpose and audience of the student lab report. We suggest, however, that it is what these two genres have in common that makes the lab report a legitimate apprenticeship genre: they share the structure of introduction, methods, results, and discussion, representing a shared way of knowing that is mirrored in other professional scientific genres, such as the conference paper, research proposal, proposal abstract and poster. (Carter, Ferzli, and Wiebe 2007, p. 294)

For science and engineering students, the activities of writing science can move from the kind of following of cookbook protocols with which they are all too familiar, to an attempt at engaging in what it means to understand, shape, and convey scientific knowledge.

Learning to Write in Laboratory Fundamentals of Biological Engineering

As a laboratory for examining the role of authentic communication tasks in the development of students' professional identities, MIT's Laboratory Fundamentals of Biological Engineering class is a particularly rich candidate (hereafter referred to BE Laboratory). The class is situated within several overlapping contexts, all shaped in large degree by the professional goals of biological engineering and all reflected in the specific writing tasks assigned.

At the first class meeting of BE Laboratory during the semester under study, students were introduced to the three-part "course mission":

1. To prepare students to be the future of Biological Engineering
2. To teach cutting edge research skill and technology through an authentic research experience
3. To inspire rigorous data analysis and its thoughtful communication. (Kuldell 2008)

Thus, from the start, students see a clear signal that their BE Laboratory experience would engage both authenticity (goal 2) and identity (goal 1) while engaging in the thinking and communicating practices central to biological engineers (goal 3) and to many other science and engineering fields as shown by the case studies featured throughout this book.

For biological engineering, preparing students for the wide range of possible roles through engagement with a variety of communication tasks is a driving force in curriculum and instruction. Biological engineering was established as a major at MIT in 2005, and at that point it was the first new major in thirty-nine years (Trafton 2008). MIT was responding to student demand: the biological engineering minor, established in 1995, was the most popular minor among MIT students (Trafton 2008). This popularity can be attributed in part to the relatively diverse nature of the field of biological engineering and the many possible directions for graduates of the program. For the twenty-three members of the first graduating class for this major, future paths reflect the many opportunities possible: roughly one third of students intended to go to graduate school to pursue a Ph.D., others to medical school, and another group to work in the biotechnology industry (Trafton 2008). These professional possibilities for graduates of the biological engineering program strongly influence the curriculum itself. As described on the program's Web site: "The [Biological Engineering] program prepares students for careers in industries ranging from pharmaceutical and biotechnology to materials synthesis, microelectronics, biomedical devices, and ecology in both basic research positions and well as project-oriented product development positions. The program also prepares students for graduate study or further professional study" (MIT Biological Engineering 2003).

In an interview at the start of the semester under examination, Natalie Kuldell, the biological engineering instructor with primary responsibility for constructing the writing tasks, spoke to her need to make learning through writing as authentic and meaningful as possible:

> One thing that's special about this class is that the students, I hope, are actually genuinely interested in communicating what they're writing about. So I think it's not just an exercise, it's not just give the teacher what the teacher is looking for, but hopefully the students have some genuine interest and some investment in the work and want to communicate about it.... It comes to this need for authenticity. And I think what's sort of special about [BE Laboratory] is this, it's not a truly authentic experience. I do not want to overstate it. Because if it were it would just be frustrating for all of us I think. But it's a semi-authentic experience where hopefully they feel like their learning matters, what they say matters, we really want to hear what they say and that this is an opportunity for them, not simply an assignment and a requirement.

The goal of authenticity permeates the work students do in BE Laboratory, whether that is the novel scientific work students pursue in laboratory or the various forms of communicating that work that students will experience.

The writing tasks students were asked to do in BE Laboratory during the semester presented in this chapter represent the broad range of writing they might encounter as professional biological engineers: an editorial rebuttal, a business plan, and a scientific research article. Students also gave two graded oral presentations: an individual talk that summarized a research article and a presentation with a peer on an original idea for a biological engineering research project.

The first two writing tasks (the editorial rebuttal and the business plan) came during the first instructional module, which involved experimental work in genome engineering, more specifically, a field known as synthetic biology. In the lab students were attempting to refactor a bacterial virus known as M13. This type of work is essential to synthetic biology, a kind of engineering approach in which biological parts are built or synthesized from their basic elements, from the nucleotides that make up DNA. The co-instructor for this module, Drew Endy, is a pioneer in this type of work, having previously refactored the bacterial virus T7 (see Chan, Kosuri, and Endy 2005), and his work with the Standard Registry of Biological Parts, essentially a user-built database of experimentally derived components for building biological systems, offered a platform for some of the writing students would do.

Thus, the laboratory work in this module was as authentic as is likely possible for a school-based context: students were working on an unknown problem (M13 has yet to be fully refactored) with a pioneer in this type of research, and the writing assignments that Kuldell and Endy designed were an effort to capitalize on the exigency of the work as well as its broad professional impact. In a sense, the writing that the students were doing would call on them to assume a range of identities—from bench researcher to editorial advocate to entrepreneur—as is necessary to join this emerging discipline. For Kuldell, the editorial response, in particular, represented an opportunity for students to get multiple sources of feedback and instruction, activities that she thought would result in the greatest learning outcomes and ones that came from the experiences of learning to write about science that she and Endy valued:

> We talked a little bit about when we each learned to write, and it certainly comes from working directly with somebody to work on it, revise it, drill very deeply into the work that you have in front of you. And students do not get a lot of opportunities for that. And so, this would be a way where this part one is first a homework assignment that we give them feedback on. Then it's exchange with their laboratory partner, who gives them further feedback. . . . So hopefully, that will be a piece they feel they've crafted. A lot of their writing, they spend a lot of time on, just hours, but it's not really crafted writing. So hopefully, this is an effort for that.

Unlike the stand-alone SciComm writing workshops described in chapter 1, BE Laboratory writing instruction was integrated into the regular routine of lecture and lab time. Students worked with a peer on their draft of the editorial rebuttal for feedback and then submitted a revised draft for feedback and initial grading to Endy and the communication instructor assigned to the class by the Writing Across the Curriculum Program. After receiving feedback and a preliminary grade, students could revise this rebuttal to improve their grades. For the business plan for the Registry of Standard Biological Parts, Endy and the communication instructor responded to the initial draft, and students could then revise that draft for an improved grade. Additional instruction was relatively informal: students could meet with the communication instructor for a conference (only one elected to do so) or could have preliminary drafts read by the teaching staff. The response on drafts from Endy and the communication instructor was in the form of a single document using Microsoft Word's insert comments and track changes features. In other words, students would see two sets of comments on their single document; in a broad sense, the communication instructor focused on the rhetorical aspects of students' writing while Endy focused on the accuracy of the scientific content and the soundness of the scientific reasoning. In actual practice, these areas would overlap a great deal. Only Endy read and graded the final drafts.

The other graded writing assignment occurred during the second of three content modules and was in the form of a research article drawing on students' laboratory work on "silencing," or controlling, gene expression in mice stem cells using the technique of RNA interference (RNAi). Initial components (introduction, methods, and figures) were assigned as homework and responded to by Kuldell, and then Kuldell and the communication instructor responded to and graded complete drafts of students' laboratory reports. Kuldell read and graded students' final versions. As was true for the first module's assignments, the instructional response was in the form of Microsoft Word's insert comments and track changes.

These three writing tasks presented a variety of challenges for the students profiled in this chapter. Although the students in the class represented a range of scientific and laboratory experiences, none had direct experience with these specific tasks. The interplay of these experiences, the students' expectations and processes based on those experiences, and the requirements of the tasks themselves revealed the dynamic between authenticity and identity in this context. In other words, although the tasks were relatively authentic to the work of professional biological engineers, students' identities as biological engineers were stretched by this range, and the process of acquiring the new discourse knowledge needed to write successfully in these contexts was challenging. Although the individual tasks could take advantage of how the students profiled identified their futures—as research scientists, military leaders, or biotech entrepreneurs—the range of these tasks challenged students to write in forms that were unfamiliar, often about content that they were not sure they had mastered, and to readers who played the dual role of professional audience and school-based evaluators.

For the student participants in this study—Kay, JoAnna, and Nedra—the overall experience of BE Laboratory was very powerful and rated high levels of satisfaction. The challenge and range of the writing tasks were significant contributors to that satisfaction, and the experiences of these students offer key lessons for conceiving of science and engineering laboratory classes as sites for students learning to write and speak as scientists and engineers.

Case Study 1: Kay—The Strengths and Limits of Experience

Of the student participants in BE Laboratory, Kay brought the most direct laboratory and scientific experience and was also the only student to state both initially and after the semester was over that pursuing an academic or research career was her goal. However, this degree of experience did not necessarily mean that her identity as a professional biological engineer was more established compared to her fellow students and that the writing tasks were relatively routine. Kay described writing as a constant challenge, remarking at the end of the semester, "I did not enjoy, no, I didn't like writing any of [the writing assignments] ... [though] I think all of them were informative and beneficial to my learning." It was not that Kay simply did not ever enjoy writing. In fact, at the start of the term, her identity as a writer was important to her, as she noted in her first interview: "I've been reconsidering lately, and I'm thinking maybe I might not even want to be a professor. Maybe I want to be a writer. I've always known I'm going to write a book at one point in my life." To pursue this interest, Kay was enrolled in a fiction-writing class in the term under study. Writing fiction or essays and writing science, however, were separate worlds to Kay, ones she was having difficulty reconciling, which added to her struggle to write in BE Laboratory.

Part of the problem for Kay was what she identified as her standard processes for learning and the reality that she would always struggle with something new. As she noted, "I do not like being thrown into something without knowing a direction or a path." These feelings are perhaps best captured by Kay's description of her writing and learning processes in her short-fiction class: "And so I've never done fiction, and I had, I struggled for like the first half of that class, like, not knowing what to do because the teacher was, like, listen, you do not want to have like an outline of where you're going, you just kind of like start writing and just it take you wherever. And I was, like, I cannot do that! I do not, you know, I need a path!" Similarly, in terms of scientific writing, Kay needed a path and struggled to write in forms with which she had not had direct previous experience. Kay had read many scientific articles and also had written in-depth technical progress reports for summer research projects, but the specific writing tasks in BE Laboratory (and their more complex rhetorical situations) were not ones she had engaged with at the depth she would need to in this class.

In terms of learning to write a scientific research article, Kay was particularly critical of previous instruction that took an immersion approach:

> What's the point, first of all, of doing something and then somebody telling you, "Well, you have to do it over again because it's not what I'm looking for," and then redoing it, and then being told again to redo it over again, rather than just somebody telling you, "Listen, this is what I expect, these are the parts and like the introduction should involve this, this, and this. The results should involve this, this, and this, and you can separate into separate parts of each part of the experiment, and do not mention any conclusions in the results, the conclusions go in the discussion...." And like just things like that that I didn't really know.

Thus, for Kay, learning to write was a struggle between the expectations and routines she brought to the tasks and instruction that could never quite be as specific as she initially wanted. Kay's lack of familiarity presented a challenge in that she sought comfort in established routines while acknowledging that she would have to undergo painful struggle while she learned something new and developed those routines. Her success as a student gave her the confidence to weather those struggles, but by no means was she satisfied with that process. As she reflected on her writing of the research report and the role of such writing in her future, she noted, "I feel like the reason that I'm just not liking it right now is because it's so new, so I'm just of kind of still learning how to do it. Once [I] kind of know what I'm doing, I will not have a problem."

The three primary writing tasks Kay produced in BE Laboratory—a rebuttal to an editorial, a mini-business plan, and a research report—were well done in a comparative sense: she received the highest overall grade in the class, and her writing was complimented at all stages by the instructional staff. These accolades were not what Kay focused on in her reflections on completing these tasks, however; rather, her focus was on the complications and sometimes painful process of getting that work done .

The editorial that Kay responded to is, in essence, summed up by the author's statement that "[synthetic biology] has a catchy new name, but anybody over 40 will recognize it as good old genetic engineering applied to more complex problems." The task for students was to draw the distinction between these two fields—genetic engineering and synthetic biology—and to present the lab research they were doing on refactoring M13 as evidence of what the promise of synthetic biology might be and the specific techniques that distinguished this work from genetic engineering.

In her first draft (see box 2.1), Kay succumbed to an issue that plagued many other students in the class: it was difficult for them to see why it was important to draw a distinction between synthetic biology and genetic engineering or what the professional niche was that Drew Endy and others had pioneered. For many, solving biological problems—no matter the approach—was the goal, and thus they did not have the broader view needed to see why carving out a distinct discipline was important strategically. In an interview, Kay said that "in the end it's all just research and, yes, we can combine different engineering aspects of it together, but if somebody thinks that it's beneficial to the field to be able to make parts and they're interchangeable, then they're going to do it, right? It's not like having a defined term for it is going to change their pushing this field forward." It is important to note that Kay offered this opinion at the end of the semester, after she had written and revised this assignment. However, she was keenly aware of the school-based nature of this task: she was writing to a reader and final grader, Endy, who was pioneering synthetic biology work and surely was convinced of the need to draw the distinction between fields.

Box 2.1
Kay's editorial rebuttal opening paragraph

First Draft	**Final Draft**
Professor Arnold of Caltech has said that "(Synthetic Biology) has a catchy new name, but anybody over 40 will recognize it as good old genetic engineering applied to more complex problems." This statement is a response to the recently highly-publicized field of synthetic biology. The difference between genetic engineering and synthetic biology can be subtle. Professor Arnold correctly recognizes that the connection between the two fields depends on existing genetic engineering methods and technologies. The concept of synthetic biology is possible because advances in genetic engineering allow us to alter genomes of living organisms. However, the key to the difference between the two fields is in what Professor Arnold refers to as the "more complex problems." Indeed, the goals of synthetic biology are challenging in the problems they present. While genetic engineering involves the transfer of genes from one organism to another, synthetic biology aims to assemble new microbial genomes from a set of standardized genetic parts (which include natural and modified genes as well as artificial genes synthesized de novo).	According to Professor Arnold of Caltech, "(Synthetic Biology) has a catchy new name, but anybody over 40 will recognize it as good old genetic engineering applied to more complex problems." Indeed, the difference between genetic engineering and synthetic biology can be subtle. However, it is important to draw a distinction between the two fields for several reasons. First, synthetic biology is a relatively new field with a unique set of goals which are distinct from those of genetic engineering. Second, there are emerging ethical and social concerns in synthetic biology which do not apply to genetic engineering. Third, because synthetic biology is a growing field, it is vital to establish a community of scientists (separate from other fields) who exchange ideas for the continued widespread applicability of synthetic biology. Finally, synthetic biology requires a modification to measurement techniques from those of standard genetic engineering. These reasons as well as direct examples of research from both fields assist in drawing the line between genetic engineering and synthetic biology.

Kay's first and final drafts of her rebuttal in box 2.1 show her awareness of this rhetorical situation. In the opening paragraph to her first draft, she characterizes the differences between synthetic biology and genetic engineering as "subtle" and does not present particularly convincing claims as to what defines these two fields. In her rewrite, however, she has retained only the opening move: the quote from the editorial she is responding to. The remaining sentences are a four-point list of concrete reasons that synthetic biology deserves to be characterized as a unique field, reasons she follows up on in the remainder of her essay. Whether she believes these distinctions or not, she strongly represents them in this task.

The second writing task for this module was to propose a mini–business plan for MIT's Registry of Standard Biological Parts, a synthetic biology database of known sequences and their functions (see http://partsregistry.org). Kay was quite familiar with the registry, having participated in the iGEM (International Genetically Engineered Machine) competition, an annual event that has student teams "design and build genetic machines" with "a library of standardized parts" (http://parts.mit.edu/wiki/index.php/IGEM:About). Nevertheless, this task was one she felt particularly unqualified to approach, contrasting her identity as a research scientist with that of a business student. "The business plan ... was the hardest thing for me to do," she explained, "just because, I think I mentioned before, I'm not really a business student, so I think I struggled with this both times, both on the rewrite and on the original."

Kay's executive summary for this task is representative of her overall text (see box 2.2). Kay changed little from the first to final draft of her executive summary, mostly by adding one sentence on the specifics about the problem that her plan would address. Unlike the editorial, this task was too far afield from what Kay knew for her to invent a rhetorical situation that would lend her some degree of comfort and address the expectations of the teaching staff. In response to Kay's revised plan, Endy commented on the lack of elements such as a pricing structure or plan for generating revenue—in other words, the business aspect of this project—noting that "this is some good writing but I'm still not convinced that you've a sound plan here. The cost of patenting parts is going to be huge, and any royalties associated from the use of parts will take a very long time to accrue." Nevertheless, Kay received a B+ on her revised draft, and she was happy to have maximized that return for what was a difficult task for her. She wrote at the end of the semester,

> If we're going to write, like, business plans, and if we're going ... to try to integrate biological engineering with business, it should be a separate class all on its own because there are a lot of things that are involved ... that we do not know about, like, how long it might take for a certain business, a start-up or something to, like, actually start generating revenue, or, like, what it takes,

Box 2.2
Kay's business plan executive summary

First draft	**Final draft**
The Registry of Standard Biological Parts aims to provide a widely-accessible interface from which researchers can design, catalog, and order interchangeable biological components and systems. The Registry aims to facilitate and expedite the research process for all sectors of the biotechnology-related field, including pharmaceuticals, agriculture, biomaterials, and university research. We strive to be the world leader in providing both a reliable source of data on biological standard parts as well as a means by which to purchase these parts in a competitively-driven market. The key to our success depends on accessibility, security of information, partnership with manufacturers, and encouragement of friendly competition amongst scientists with the goal of facilitating advancement in scientific and technological breakthroughs. We are dedicated to the cause of synthetic biology in making engineering of novel biological systems reliable, efficient, and innovative.	The Registry of Standard Biological Parts aims to provide a widely-accessible interface from which researchers can order catalogued, interchangeable biological parts and to which they can contribute novel parts and devices. The Registry aims to expedite research for all sectors of the biotechnology-related field, including pharmaceuticals, agriculture, biomaterials, and university research. We strive to be the world leader in providing a database to store standardized biological parts, ensure a reliable source of information on part functionality, and sell parts as a product in a competitively-driven market. We plan to address current problems in the Registry which include the unreliability of parts, competition from alternative markets lacking standardization, and the present absence of revenue from distributed parts. The key to our success depends on accessibility to our products via thorough documentation, security of information and patent options for part designers, and encouragement of friendly competition amongst scientists with the goal of facilitating advancement in scientific and technological breakthroughs. We are dedicated to the cause of synthetic biology in making engineering of novel biological systems reliable, efficient, and integrative.

like, what resources you need.... We do not really know that, and that's not something you learn by sitting and doing transfections in lab.

The business plan might be an authentic task for biological engineers in establishing funding and longevity for their field, but for students such as Kay, it was too far removed from her expectations and from what she thought was reasonable for a laboratory class and from her identity as a student scientist. Ultimately she was not opposed to the idea of having to write these plans, but she thought that the stakes and time required were too high for what she felt she had learned: "I feel like it could have been a homework assignment, [and] we could have just gotten just as much out of it.... It wouldn't have been such a, like, a huge ordeal, and ... what I learned from this would have been the same."

Kay had the most amount of familiarity with the third writing task: a report on the research she was doing in lab on short interfering RNA (siRNA) as a technique to silence gene expression, one in which data analysis would be most important. Kay liked the fact that this task was broken into several subtasks as homework assignments; thus, she was able to get feedback on her introduction, methods, and figures before having to submit the complete report. Kay's experiences with writing up original science, while limited, offered a complication as well as a complement to her process. Because of the timing of the instruction, students' introductions were due before they had finished compiling results in the lab, and Kay expressed her frustration with this timing: "I didn't like the fact that we had to write the introduction before we even, like, finished all the experiments, right? Because it was just, for me, like, I need ... to know where I'm going, so it's been done.... When I was writing, it was really awkward saying, 'We did this' when we hadn't even done it yet." Thus, although the lab report was in many ways an authentic task that offered students the opportunity to write up original results, the timing of that task could not quite mirror that authenticity, given the timing of the semester and the need to break up the task as a series of homework assignments. In a sense, Kay found it difficult to craft the story of her research, given the forced nature of its timing.

In terms of Kay's writing, her report abstract offers an example of the strengths and struggles she brought to this task (see box 2.3). Her first draft conforms to all of the expectations for what a scientific abstract should contain (essentially, a mini-version of her entire report), but as Natalie Kuldell's comments point out, Kay did not give much emphasis to her "key data" and left out specifics of her results. In her final version, Kay addresses those concerns with lots of specific data (essentially four of the last five sentences), as well as a concluding sentence that nicely mirrors the conclusion to her report. Overall, Kay's report is a strong example of a complete abstract for this lab work, and she received a 95/100 on her final report.

Box 2.3
Kay's lab report abstract

<table>
<tr>
<td>

Draft with instructor's comments and edits (in italics)

Short interfering RNA (siRNA) has been shown to silence target gene expression. However, successful silencing and off-target effects by siRNA are often difficult to predict. We designed siRNA*s* to target six 156 base pair regions of the Renilla luciferase gene and measured their effectiveness at silencing luciferase expression as well as *their* effect on the expression of other genes in mouse embryonic stem cells transfected with plasmids containing Renilla and firefly luciferase genes. The firefly luciferase was not targeted by the siRNA and was not shown to decrease in expression *('was shown not to decrease'?).* Downregulation of Renilla luciferase was observed for some, but not all of the siRNA designs. Differential gene expression *(say how examined)* was also observed in MES cells transfected with the designed siRNA, which indicates off-target effects. *This abstract is a clear description of the goal though you make it esp hard on yourself in that you try to describe the class data for all the siRNAs and move between "the siRNA" and the class pool. You do give some of the key data you collected around luciferase activity but the off-target effects of the siRNA need more detailing—even if you didn't find a single pathway that was affected you can still describe the number of genes you found significantly changed and by how much. Non-specific effects should also be considered.*

</td>
<td>

Final draft

Short interfering RNA (siRNA) has been shown to silence target gene expression. Such control of cellular transcript levels has many applications, such as down-regulating over-expressed genes in diseased cells. However, successful silencing and unintended off-target effects by siRNA are difficult to predict. We designed siRNA to target the 469–624 base pair region in the *Renilla* luciferase gene. We then measured the decrease in luciferase activity upon addition of our siRNA to mouse embryonic stem (MES) cells harboring plasmids with *Renilla* and firefly luciferase genes. To detect off-target effects, we measured the levels of messenger RNA (mRNA) upon addition of our siRNA in a mouse whole genome microarray. Results showed a 3-fold as well as a 1.1-fold downregulation of *Renilla* luciferase with our siRNA in duplicate luciferase assays. The firefly luciferase was not targeted by our siRNA and protein activity did not decrease. Microarray analysis showed differential gene expression in MES cells transfected with siRNA, indicating off-target effects. No single gene families or pathways were particularly affected nor were there non-specific effects amongst the affected genes. A total of 8,031 locations on the microarray had a 2-fold induction of mRNA, while 29,396 sites had a 2-fold reduction. We conclude that our siRNA is not optimal for targeting the *Renilla* luciferase gene in MES cells due to minimal reduction of *Renilla* luciferase activity and a large number of off-target effects.

</td>
</tr>
</table>

In reflecting on her process for this report, Kay noted that her preference was for writing the introduction and the discussion because of the opportunities they provided to engage in the thinking processes of "real science." As she noted, "I like thinking about ... discussions ... because you kind of interpret.... It's real science; it's like you're trying to make a hypothesis about why this result, you know, maybe it wasn't expected, but you can sort of learn new things from that."

Overall, although Kay had very defined expectations for what a laboratory class should entail, expectations that were quite disrupted by the variety of discursive tasks in BE Laboratory, she reported that her BE Laboratory experience was extremely positive. In an interview, she called the class "amazing.... Everything you could possibly put into a class you did, like you have to work, you had to learn how to work with others, you have to learn how to make a business out of biology, you have to know how to write papers, how to read papers, how to make figures, like how to do experiments in lab, like, everything you can do was [in] this class."

As far as the class's contribution to her growing professional identity (and her discursive identity as a professional engaged in the communication tasks specific to her field), Kay was also positive, noting at the end of the semester that the class "gave me an idea of ... the choices that I have in my future, whichever track I choose to take." In a follow-up e-mail interview near the end of the semester following the one in which she took BE Laboratory, Kay was even more positive about the authentic nature of her BE Laboratory experience and its benefits to the course work she was doing as a BE major and her future as an academic researcher. She wrote, "I think there's a very close relationship between what was expected in [BE Laboratory] and my professional goals. Communication is key in any profession but especially if I want to be a professor in a field of science. The sort of writing and presentations I gave in [BE Laboratory] are going to be part of the job."

As a student experienced in the processes and practices of research science, Kay challenged notions of authenticity. An authentic task for a professional scientist or engineer often requires knowledge and experience that students do not necessarily have, particularly when those tasks reach broadly into the work of a professional, such as the business plan in BE Laboratory. Or the professional knowledge that an experienced student might have, such as Kay's experience with the process of writing up lab results, might be in conflict with the time schedule and due dates of the academic semester. Perhaps it will not be until graduate school—as shown in the case studies in chapter 3—that Kay will best balance developing a professional identity and engaging in the authentic tasks of biological engineer. Nevertheless, her overall positive experience attests to the need to make professional activities and standards the driving force in curriculum and teaching.

Case Study 2—JoAnna: Short-Term Sacrifice for Long-Term Results

While Kay's experiences with scientific research and communication gave her a foundation on which to build and a set of expectations that at times were challenged by the work she did in BE Laboratory, for JoAnna the science and the writing were largely unfamiliar and were not necessarily intended to lead to a career as a researcher. As a junior enrolled in ROTC and from a military family, JoAnna's intent was to serve in the U.S. Navy following graduation. She chose to major in biological engineering because she believed such study "would make me stronger intellectually." For her military career, JoAnna believed that biological engineering was a good choice to give her a well-rounded knowledge and skill base: "I'm not going to use my biological engineering degree directly. But, you know, the principles I learn, and the ability to study and work hard and memorize and problem solve, that will all serve me very well."

JoAnna's ability to endure short-term difficulties for long-term gains was consistent with her approach to choosing a major and a career and to her overall discipline and work ethic. In a sense, the authenticity of the communication tasks of BE Laboratory was secondary to the overall skills and strategies she would learn from them. In other words, it was not the concrete nature of the scientific experiments or writing strategies that were important to her but the ways she might abstract from those tasks and apply them to her future work, whatever that might be. This kind of long-term thinking seemed key to JoAnna's identity, one in which the idea of sacrifice for one's country and the importance of larger goals were vital. At the start of the semester, she said, "I'm really excited to be in the Navy. I'll get to be a leader and impact a number of people's lives very directly, and then in a more intangible way, protect my country and everything it stands for."

BE Laboratory was JoAnna's first significant laboratory and first experience with writing up scientific content, but she did bring extensive experience as a journalist, having been managing editor of her high school newspaper. Once at MIT, however, JoAnna felt that her writing suffered as a result of few opportunities to write, particularly to write in scientific and technical forms:

> When I was a senior in high school, like, I did theology and English and history and all sorts of classes that you had to write a lot, and you had to write a lot frequently. And here, I come to MIT, and I'm doing problem solving all the time, P[roblem] sets. And so when I try and return to writing, it's not as familiar. I mean, it's not like I'm producing a kindergarten-level paper, but it's just not the same product that I could produce as a senior in high school. And it's frustrating.

To overcome these frustrations, JoAnna distinguished herself by seeking all available opportunities for writing assistance, conferencing with the communication instructor assigned to the course for each of her tasks (the only student to do so). As she reflected on these tasks at the end of the semester, she remarked that she was unsure of her initial approach and relied on instruction and feedback to improve. It was a process she anticipated as painstaking, but one she believed she needed to undergo to build foundational skills for future communication experiences.

In terms of her specific work, JoAnna faced the same dilemma that Kay and many other students did for her editorial rebuttal: not being convinced themselves that a distinction needed to be drawn between genetic engineering and synthetic biology. The introduction to her first draft (box 2.4) argues by analogy to *Harry Potter* that both genetic engineers and synthetic biologists are today's "witches and wizards." The distinction she does draw between the genetic engineer's "science mindset" and synthetic biologist's "engineering mindset" does not seem particularly distinct or developed, and by her last focusing sentence, she seems most interested in the common goals of these two fields. Her rewrite does a bit more to draw a distinction, but the focus is on the "controversy" of the distinction, and the final sentence makes an even stronger argument that the two fields are aligned in the kinds of problems they attempt to solve and the "related methods" they use to solve them. The difference between them, then, are the "attitudes."

In reflecting on this task at the end of the term, JoAnna was still not convinced that drawing a distinction between these two fields was important. She noted, "No. It's not important to make the distinction. I mean, I'm sure you had synthetic biologists and biological engineers who are way up there like professors and have tenure or whatever and they are like, oh, my job is vastly different from the synthetic biologists, but it all boils down to making a difference and that's the most important thing."

The long-term goal was again what was important to JoAnna: "making a difference," no matter the approaches or "attitudes" of those trying to achieve that goal. In her end-of-term interview, JoAnna also tied her beliefs in this assignment to her identity as neither a synthetic biologist nor a genetic engineer: "Maybe part of the reason [for not drawing a distinction] is that I have no desire to be either. . . . I'm studying about them and what they do is incredibly interesting, and I'm glad I chose this major, but at the same time I'm more able to take a step back and say, well, that's not really the most important thing." What might be an authentic task to biological engineers, then, was not one for JoAnna, given her future plans.

In terms of her business plan (box 2.5), JoAnna's executive summary first draft did contain fairly specific plans on refocusing the intent of the Registry of Standard Biological Parts, but like many other students, she had difficulty figuring out a specific audience for this document and thus the level of detail to offer about the registry itself. She resolved this problem to a large degree in her revised draft, which contained much more preliminary information to define the registry and specifically set out her intent

Box 2.4
JoAnna's editorial rebuttal opening paragraph

First draft	Final draft
Through her *Harry Potter* books, J. K. Rowling created an imaginative world where witches and wizards instantaneously replaced fish gills with human lungs and other such feats. Today's witches and wizards are much slower, but work on tasks that seem just as impossible: they are synthetic biologists and genetic engineers using the magic of science and engineering to solve real-life challenges. Frances H. Arnold, a professor at California Institute of Technology, claims that synthetic biology is a catchier version of genetic engineering applied to more complex problems. While both use bacterial transformations and yeast transfigurations, their methods and mindsets differ; genetic engineers use a science mindset to focus on inserting already made genes while synthetic biologists use an engineering mindset to create genes from scratch and work to make them more functional. Though their methods may differ, both are working to solve the same issues in medicine, environment and biology.	Through her *Harry Potter* books, J. K. Rowling created an imaginative world where witches and wizards instantaneously replace fish gills with human lungs and other such feats. Today's witches and wizards work slower, but their tasks seem as impossible: they are synthetic biologists and genetic engineers using the magic of science and engineering to solve real-life challenges. Though scientists and engineers can agree that their work is exciting, developing these two disciplines as separate fields has sparked controversy. Although Frances H. Arnold, a California Institute of Technology professor, claims that synthetic biology is a catchier version of genetic engineering applied to more complex problems, the two have significant differences. While both use bacterial transformations and yeast transfigurations, their methodologies differ; genetic engineers use a science mindset to focus on inserting already made genes while synthetic biologists use an engineering mindset to create genes from scratch and work to make them more functional. These scientists and engineers tackle identical medical, environmental and biological issues with related methods but different attitudes.

Box 2.5
JoAnna's business plan executive summary

First draft	Final draft
The Registry of Standard Biological Parts will provide a wealth of information and consulting services to synthetic biologists. One component of the Registry is the database of BioBricks, standard biological parts that can be used to create new pathways, bioengineer a new organism or design and build a living machine. This database of information allows scientists and engineers to use previously existing genetic sequences in innovative ways to solve complex problems in medicine, biology or the environment. The second part of the Registry will be the consulting services; Registry employees will venture into synthetic biology labs where BioBricks are used. They will provide engineers and scientists novel methods and techniques and give general advice about the experiments. Due to recent technological advances, synthetic biology is the fastest growing field in science or engineering. The re-designed Registry will provide better services, thus facilitating future advances in the field while becoming a lucrative business. Motivated employees are the keys to success in any business; the Registry is no different. The exciting synthetic biology advances the Registry facilitates will inspire the employees to be enthusiastic about their work. That enthusiasm will lead to the best product possible and will transfer to the entire field itself.	The Registry of Standard Biological Parts is a database of BioBrick parts, pieces of DNA that can be used to program living organisms. The hope of this redirection is to mold a company that provides a wealth of reliable information and consulting services. Theoretically, the BioBrick parts already on the website can be used to create new pathways, bioengineer a new organism or design and build a living machine. Unfortunately, many of them are unreliable and do not function as advertised. This database of information should allow scientists and engineers to use previously existing genetic sequences in innovative ways to solve complex problems in medicine, biology or the environment. With this new technology, a guide will assist in product development. The second part of the Registry will provide that map though the consulting services; Registry employees will venture into labs where BioBrick parts are used. They will assist engineers and scientists, teaching novel methods and techniques and giving general advice about the experiments. Due to recent technological advances, these fields are the fastest growing in science or engineering. The re-designed Registry will provide better services, thus facilitating future advances while becoming a lucrative business. Motivated employees are the keys to success in any business; the Registry is no different. The exciting advances the Registry facilitates will inspire the employees to be enthusiastic about their work. That enthusiasm will lead to the best product possible and will transfer to the entire field itself.

in this document ("The hope of this redirection is to mold a company that provides a wealth of reliable information and consulting services").

As she reflected on her writing of this task, JoAnna did not find it difficult to take on the identity of a biological engineer as required for this task:

> I thought it was easy, just because of we'd been working in labs so long as a biological engineer.... All of my classes are, like, I'm doing biological engineering work a lot, between going to class and studying for class and doing my p[roblem]-sets and stuff. So imagining myself wasn't that challenging, because I feel like I'm doing that work already. It's just imagining myself with a lot more information inside of my brain, was the only difference.

JoAnna's ability to imagine a professional identity made the task itself authentic and meaningful. Given her level of comfort, she found success in this task, achieving one of the highest grades in the class.

The laboratory report presented a variety of challenges for JoAnna. She described the writing and revision process most comfortable to her as a relatively linear one: start at the beginning and write to the end of a document, and then revise in a similar fashion. However, the laboratory report was broken up into a series of homework assignments that called for presenting different parts of the whole task in an order different from their final presentation, and JoAnna also found herself composing in a new way, one that was far more circular in its revisiting of different parts of the whole than her previous linear approaches: "I think it was just that first, like, I do not even know what I'm doing, I've never done this before. And then I started getting into it.... I finished the results and then immediately went back and revised my introduction, and then it just felt more circular than linear to get the final product." Additionally, the individual tasks were new to JoAnna and markedly different from the laboratory reports she had written in high school—ones that were focused on the questions, in her words, "what did you do, and how did you like it, and what did you learn from it?"

In the first draft of her laboratory report abstract (box 2.6), JoAnna offered the essential moves of focusing on the particular line of research she explored and justifying that intent, but like many other students, she did not offer very much in terms of her specific results. As Kuldell's comments indicate, JoAnna also needed to clarify a scientific point about what she actually did. In a sense, Kuldell offered feedback to move JoAnna to write a more authentic scientific abstract, as she did for all of the students. Kuldell's experience as a scientist and scientific writer offers an apprenticeship opportunity for JoAnna, and in her final draft JoAnna was able to incorporate those necessary elements. Overall, she received a 92 on her final draft, one of the higher grades in the class.

At the end of the semester, JoAnna reported a high level of satisfaction with her BE Laboratory experience, describing it as follows:

Box 2.6
JoAnna's laboratory report abstract

First draft with comments	**Final draft**
~~Studies have indicated that~~ *(suggestion for wording)* Short interfering RNA molecules (siRNA) can regulate gene expression by inducing mRNA degradation through a process called RNA interference. Currently, siRNA technology is being used to develop therapeutics and study gene function in vitro and in vivo. Though researchers can transfect short interfering RNA molecules (siRNA) into cells to effectively quench gene expression, successful siRNA design methods are still being developed as many designed molecules do not produce expected results. Following specific guidelines, researchers *(we?)* designed a 21-nt siRNA molecule that effectively degraded Renilla Luciferase mRNA molecules in mouse embryonic stem cell *(so a technical point here is that you examined luciferase protein with your luciferase assay and from a decrease in enzyme activity and an understanding of RNAi you infer a decrease in mRNA)*; however, the siRNA molecules produced undesired effects, unintentionally altering mouse gene expression. *(specify in more detail how you know this and what unintended effects you found).*	Short interfering RNA molecules (siRNA) can regulate gene expression by inducing mRNA degradation through a process called RNA interference. Currently, siRNA technology is being used to develop therapeutics and study gene function in vitro and in vivo. Though researchers can transfect short interfering RNA molecules (siRNA) into cells to effectively quench gene expression, successful siRNA design methods are still being developed as many designed molecules do not produce expected results. Following specific guidelines, we designed a 21-nt siRNA molecule to target Renilla luciferase. After examining enzymatic activity in mouse embryonic stem cells with a luciferase assay, we concluded that the siRNA effectively quenched Renilla luciferase activity. Microarray data showed smaller overall transcription rates in MES cells transfected with experimental siRNA as compared to the rates in cells transfected with an siRNA molecule known to quench Renilla luciferase activity. Nevertheless, the siRNA's successful quenching of Renilla luciferase illustrated the success of our design principles.

> Phenomenal. I absolutely loved it. It was challenging, I mean, I'm not going to lie. It was a lot of work and it was hard but I think we wrote so many different assignments that like, I mean, I've never written a business proposal, never written a scientific lab report. I mean, I'd written like essays analyzing like editorials before but it was different because there was more science involved and it stretched me a lot and I feel like, I mean, I improved a lot by the end.

By the end of the following term, JoAnna's feelings about the value of the class were still strong. She called it "a fantastic experience. The amount and certain papers were daunting, but overall the experience was a great one. The class was taught well and the challenges pushed me in a constructive manner." And she still intended to pursue a military career. While she was doing some writing in her classes that term, as was true for much of her learning, she could not see immediate applications to her professional career, but she could distill long-term value: "The writing I have been doing is very different from what I will be doing in the Navy, but any practice that hones one's skills is vitally important." Thus, for JoAnna the notion of authenticity was stretched to include not merely the genres that might or might not map on to future professional tasks, but the skills and strategies—the learning experiences—that one takes from engaging in those tasks.

Case Study 3: Nedra—Struggling with Time

Nedra, the only sophomore among the participants in BE Laboratory (the other two were juniors), had had a fair amount of research experience, having completed two undergraduate research projects in biological engineering labs closely related to the work she would do in BE Laboratory. The most powerful factor for Nedra to overcome was the constraint of time she had to complete class tasks. In a sense, Nedra's identity as a writer, particularly a scientific writer, was far from positive at the start of the semester and did not seem to improve much as a result of her BE Laboratory experience (though she ranked the overall experience quite highly). The interplay between an identity as a weak writer and the constraints of time and due dates necessarily created by writing for class assignments did not work in Nedra's favor. Her grades on her writing assignments were consistently lowered because she handed in her work late, and even communication activities that she thought went well—her final oral presentation conducted with a partner—followed last-minute cramming and lost points as a result of running past the allotted time. Time was not Nedra's ally in this class.

Throughout the term and in follow-up correspondence, Nedra consistently identified her need to be a strong communicator for her chosen career goal, a career in business. After obtaining her undergraduate degree, Nedra imagined that "I think what I'm going to do is I'll probably just work hard in whatever job it is just to get some money to go to business school. But after business school I want to either participate in a start-up or do something with biological engineering.'' Nedra saw her strength in oral communication as particularly well suited to a career in business, though she professed to have a strong desire to do scientific research and believed she needed a strong scientific background if she were going to succeed in business—the reason she chose biological engineering as a major.

Nedra recognized that writing was not a strength for her, though she had received accolades for her writing. For Nedra, those awards were relative to the community. Compared to her high school cohort, she felt she was a good writer, but at MIT, she felt at a distinct disadvantage, noting in an interview that ''my writing background is relatively worse than a lot of other MIT students.'' In trying to explain her unease with writing, Nedra drew on an analogy to her relative mastery of chemistry and the implicit knowledge needed for that mastery: ''When I was in high school, I met some people who were in chemistry, and they were given the equations, and they were given a methodological way of approaching it, but they just couldn't do it. And I'm like, I do not understand why you cannot do it. You start with something, you end with something. It's not very hard. And they're like, I just do not get it. So I haven't just figured [writing] out or it's just not working somehow.''

When it came to the writing tasks for BE Laboratory, Nedra's performance was in many ways determined by the time constraints she faced as a sophomore still adapting to the academic pace of MIT, trying to find balance of classes, research jobs, and the expectation for MIT students to be, as she described, ''saving the world.'' Perhaps her lack of confidence in writing led to no small degree of procrastination as she faced the writing tasks due in BE Laboratory; however, even her opportunities to revise were challenged by the time she was able to make available. In fact, for her first two tasks, the editorial rebuttal and the business plan, she was not able to find time to revise at all, and for the third task, the laboratory report on siRNA, she handed her initial draft in late and then ran out of time in the revision process, and, as a result, her rewrite suffered in terms of final grade.

The editorial that Nedra chose to respond to questions if the potential for synthetic biology to build organisms from the ground up is actually ''creating life.'' In the first paragraph of her rebuttal (box 2.7), Nedra succumbed to two problems: not communicating clearly the point of the editorial she is responding to and not making her stance on the issue clear. The reader does not quite know what to make of the ''huge crowd'' in her opening sentence, who ''protests in an uproar,'' or the logic of, ''Because those cells were dependent on a rich culture medium, the author concluded that thinking in a solely scientific way was not enough,'' or the position that ''research should continue

Box 2.7
Nedra's editorial rebuttal opening paragraph

First and final draft

Now that we are getting closer to determining how we, as humans, can modify life, a huge crowd protests in an uproar. Are we allowed to redefine life? Should we continue thinking of this as a scientific concept, or is it out of our hands? In a recent editorial titled, *The Meanings of 'Life'*, the author stated that, "It would be a service to more than synthetic biology if we might now be permitted to dismiss the idea that life is a precise scientific concept." This comment was stated in reference to scientists of the Venter Institute who tried to artificially create bacterial cells. Because those cells were dependent on a rich culture medium, the author concluded that thinking in a solely scientific way was not enough. He stated that we, especially synthetic biologists, had to acknowledge the "divine spark" that gave life to us all. In reality, we shouldn't be dismissing the idea that life is a scientific concept. Rather, the synthetic biologists doing this research and other research should continue because we *can* believe that life is a scientific concept, especially in synthetic biology.

because we *can* believe that life is a scientific concept, especially in synthetic biology." In her reflections on this assignment, Nedra acknowledged the problem: "I mentioned before that sometimes my writing is a hit or miss. And this was one of those misses because I actually didn't address the topic the way I was supposed to so it's kind of one of those default zeros, like, well you didn't even read the question right, so big X."

Although students had a choice of several editorials to respond to for the assignment, most selected the one that Kay and JoAnna did on genetic engineering versus synthetic biology. Nedra's choice of this particular one was reflective of her creativity as a student and her desire to take risks, but she also suffered the results of those risks: "I didn't want to do the second topic, because it was just comparing two things.... So I took the harder one and paid the consequences."

For her business plan, Nedra felt a degree of comfort given her orientation toward a business career. Nevertheless, time was a factor again: "I think I started around two in the morning so it wasn't as fun as it was supposed to but I had a test that next day as well, so it was a little difficult." Working all night on a draft that was due the next day, Nedra tried to follow a process that generally worked for her, but it was not necessarily one that would work for her reader:

It wasn't very organized; I was kind of all over the place. And I tried to group things into themes, because when I plan out my essay, I always plot, I wrote out all my ideas on a sheet of paper and they were all over the place. And then I go back and I figure out how I can group them into

Box 2.8
Nedra's business plan executive summary

First and final draft

The Registry for Standard Biological Parts (RSBP) has come very far with respect to increasing its database and evolving on its own after its recent branch off from MIT. It has future plans to expand and provide more services to attract more customers. We intend to standardize all parts and have a thorough evaluation and documentation of each part so they can be properly used. All parts will be able to be purchased from a catalogue in the future and will be delivered to your door soon after. No longer will users have to worry about the hassle of finding a place for a sequence to be synthesized or worry whether the sort of sequence has a likely chance of working. RSBP, over the next 10 years intends to evolve into a more central "one stop" manufacturing company where a specific description of a series of parts can be dropped off and a fully developed sequence will be mailed to its customers' door. Moreover, new more stringent procedures for adding and purchasing parts will be enforced in order to uphold the necessity for ethical liabilities. With its new partnership with the government, we at the Registry for Standard Biological are proud to take part in the starting of synthetic biology, a new science that is rapidly gaining popularity. We strive to become the reliable, hardworking company that its customers have always deserved.

paragraphs and under main ideas. But the thing is that sometimes one topic could fall under one or the other. And obviously the person who is reading it has a certain opinion about it and saying, oh no, no, I thought this should have been in this paragraph instead of this paragraph.

In her interviews, Nedra consistently represented rhetorical situations in this way, as a matter of opinion over choices such as organization or evidence. What felt right to her as she was writing might not be perceived as the best approach to her reader, but she would not know that until her reader offered a response. And because she could not create time to revise based on that response, her writing product suffered. As she noted about the business plan, "One of the comments was I didn't have enough detail. I mean, I didn't really know how much detail to put in. I wanted to be concise. But I guess I was too concise."

In terms of her plan itself, her executive summary (box 2.8) was marked by a difficulty many other students faced in terms of what constitutes a business plan. Rather than contain concrete methods for developing the registry and a justification for those methods (addressed to an audience that might potentially fund this plan), her approach is more of a marketing angle or an in-house document meant to increase employee morale. Nedra recognized this mismatch between instructional staff expectations and what she wrote and thought it could have been addressed with more concrete models of what she should write: "Because I feel I work best by example, and since we didn't have any example, it was more like, write up a business plan. Write whatever, do not worry about it. And then they're like, yeah, that's not what we wanted."

The laboratory report was a relatively new task for Nedra. Given her perceptions of herself as a poor writer compared to her peers and her lack of experience with scientific writing, she expressed some anxiety in an interview conducted before she wrote her report: "For this lab report I think what I'll focus more on is actually mimicking an actual lab report. I've never written a lab report before so it's going to be interesting.... I've never really written in a scientific way. And I noticed one of the comments that my TA gave me was that you need to, I write like I talk. So it's very informal and I have to write more formal for the lab report. So it's probably going to be a struggle for me."

Once she was finished with her report, Nedra, like Kay and JoAnna, was glad to have been able to complete portions of it as homework assignments and get feedback on those. However, for the complete first draft of her report, time again became a problem for Nedra. She reported at the end of the semester that she "took a lot longer than other people to complete the project. I took some extra days. It was hard. I hit a roadblock at one point,... and I didn't know how to write it up, and so I think that was the hardest part, getting past that roadblock."

As was true for her other writing assignments, Nedra reported having difficulty figuring out her readers' expectations and what forms would best fit her final product, particularly for the methods section:

> What was hard for me [was that] they told us to refer to articles, but they didn't exactly want it in article form. Like, that's what they told us but they wanted more about how we did things, whereas in normal articles they leave out all those details.... And I feel like our instructions should have been a little more clearer about what we should have done and what was different from whether it was an article or just a report, like we did in Module One. So I had a little trouble with that. A lot of trouble with that.

Nedra's revising process after receiving feedback on her initial draft was again largely affected by the time she had available to revise. Her approach was to start at the top with her abstract and respond to comments until she ran out of time. But time ran out before she got very far:

> I really wanted to work on my rewrite; one, because I got a horrible grade and I wanted to bring it up as much as possible. And two, because I was really frustrated and I definitely felt like I could have done better if I had more time. So I definitely went in order, which in retrospect was a really bad idea.... I rewrote in order from whatever it was on the first page to the end. And I didn't get very far.... I ran out of time. And so I didn't get many points back on my rewrite.

Nedra's abstract to her lab report did much of the work of an abstract, but as Kuldell's comments indicated, she started too generally, did not focus on the purpose of her actual investigation, and did not report results completely (box 2.9). It is important to note, too, that Kuldell's comments acted to assert professional competency, whether feedback on what an abstract should contain or specifics of professional language, such as using "we" rather than, "My partner and I." Nedra addresses some of these concerns, but not in a deep sense or not in the sense that an abstract written by a scientist would appear.

Despite her trials and tribulations of writing in BE Laboratory, by the end of the semester, Nedra viewed the experience as largely positive, particularly the emphasis on independent research. As she described, "In [BE Laboratory] we were told to do things on our own: decide our own SRNA strand, refract our own genome, come up with our own ideas, alter the methods and report them back. And having that sort of control made me really happy. I was very passionate about it and I really took it seriously where we were doing our own things. And that was fun." Nevertheless, Nedra recognized that authenticity was elusive given the time line of a semester and her experience with scientific research. She noted, "I feel like in the real world you get a little longer of a timeline for coming up with an idea to refactor a genome or writing up a lab report or coming up with a research proposal. So I mean, I feel like I'd like real life better, but this was kind of real life or your entire research career compacted into one semester. So it was good in some ways, but I feel like I came out of the course with a lot of gained insight."

When contacted near the end of the semester after her BE Laboratory experience, Nedra was still positive about what had occurred, but in perhaps unexpected ways. She had decided to change her major to mechanical engineering and credits her BE Laboratory experience with helping her make that decision: "I think [BE Laboratory] helped me realize what my interests were, which caused me to change my major." She also noted that her writing experiences in BE Laboratory helped her feel "more easy with doing writing now (but I do not do much writing at this point)." Thus, as was also true for Carla in SciComm, a potential effect of authentic professional experiences is to help students make decisions about their future—ones that may include entirely different professions.

BE Laboratory also challenged Nedra's established student routines, particularly her time management skills, and these complications speak to a certain developmental readiness for students to learn from authentic scientific writing tasks, particular for students who self-identify as struggling writers, as Nedra did. Unlike April in SciComm, who was trying to master the scientific content, Nedra was struggling to get work

Box 2.9
Nedra's laboratory report abstract

First draft with instructor comments	**Final draft**
RNA-interference is a newfound technology useful for the degradation of mRNA. This *(what is "this"?—newfound technology?)* happens by introducing double stranded RNA into the cell and letting the cell cut these strands into small interfering RNA (siRNA) → *I recommend reconsidering this introduction info since it is probably too general for the abstract, and in your experiment you directly added the siRNAs.* If the siRNA find a homologous mRNA sequence, it is able to degrade this mRNA and thus inhibit the expression of that gene. My partner and I *"we" (wording suggestion)* were interested in finding a way to silence the *Renilla (italics for latin)* luciferase gene by RNAi, specifically by siRNA *(is this really the purpose of your investigation? You start the abstract by describing RNAi as a newfound technology and so the application of this technology to silence a non-natural gene in mouse cells seems too quirky of an experiment for a reader or reviewer to want to read more).* To do this we first *(avoid temporal markers like "first" or "then" since they do not add clarity. The order you present things will let the reader presume an order for importance and procedures)* designed our own siRNA sequence according to general siRNA designing guidelines. We then tried this and other siRNA sequences out by lipofecting cells *(what kind?)* with the psiCHECK2 plasmid *(the reader will not know what this is)* the siRNA sequences. The siRNA sequences used were experimental (designed by us), scrambled (recommended by Dharmacon), and a validated	New discoveries show that short non-coding double stranded RNAs, like short interfering RNA (siRNA) can regulate gene expression on the post-transcriptional level. siRNA is one of several RNA-dependent regulatory methods known as RNA interference (RNAi) where siRNA silences the expression of homologous sequences in the cell it is introduced to by degrading them. We were interested using siRNA to silence the *Renilla* luciferase gene, a gene that causes its host cell to luminesce. We introduced a vector which expresses the *Renilla* and firefly luciferase gene into mouse embryonic stem cells to easily assess the effectiveness of the siRNA. To do this we designed our own siRNA sequence according to general siRNA designing guidelines and lipofected this siRNA and two control siRNA, a validated and scrambled strand, into mouse stem cells. After growing the cells for two days, we performed a luciferase assay to measure whether we were able to silence the expression of the *Renilla* luciferase gene and found that both the validated and the experimental siRNA were able to silence the expression. Pictures from a microscope verify that the mouse stem cells had no noticeable phenotypic effects cause by any of the siRNA sequences. Moreover, through the analysis of a microarray, we were able to determine that only 119 of the 44,000 genes had a log 2 ratio value greater than 2 or less than −2. This simply tells us that there was a difference in the off-target effects of these two siRNA strands.

Box 2.9
(continued)

(recommended by Promega) *(this level of detail about controls is better suited for other parts of your paper)*. After growing the cells for two days, we performed a luciferase assay. Our results showed that both the validated and the experimental siRNA sequences were able to silence the expression of the luciferase gene. Moreover, through the analysis of a microarray, we were able to verify that the mouse stem cell was not being significantly affected by these siRNA strands *(not sure what "strands" means)* relative to each other *(the siRNAs were relative to each other?) You need to expand on this section of your work as well as growth of cells. Were there any general effects you could see? How many genes were affected and by how much? Non-specific or off-target effects evident? Even if the answer is no, that must be explicitly described in the abstract.*	

done within the many demands on her time as a student at MIT. She was a sophomore at the time of this study, as compared to students with an additional year of student experience and, presumably, with additional experience with time management.

Summary of Laboratory Fundamentals of Biological Engineering

The case studies presented in this chapter indicate both the importance of authentic, professional writing tasks as key curricular components and the challenges for students and instructional staff in working with such tasks. The relationship between authentic tasks and students' developing professional identities, the need for sound writing instruction, and the long-term outcomes of writing in classes such as BE Laboratory are key lessons.

The relationship between students' futures and their engagement with authentic communication tasks was strong. Kay and JoAnna's senses of their futures was confirmed by engaging in the range of writing tasks they faced in BE Laboratory. This range pushed them to grow and learn, and they were confident that in the future, they would apply what they had learned, whether it was for Kay's career as a scientist or JoAnna's as a military leader. For Nedra, the authentic nature of the tasks confirmed that biological engineering would not suit her future career, and she moved on to a new major. Thus, for the three participants, the authenticity of the tasks fed tangibly into the identities they currently held and imagined they would hold.

Students develop as writers through engaging in a range of professional tasks and from feedback from a range of potential readers. For students in BE Laboratory, the range of professional tasks pushed them beyond what might have been comfortable or routine, but this extension was supported with extensive feedback from a variety of sources, peer and professional, and an iterative writing process. In a sense, students' development of identities as writers with successful routines and strategies needed explicit attention—whether direct instruction or reflective practice. These are fundamental practices of writing classrooms but are not always present in disciplinary classrooms and laboratories, given concerns about losing time to cover content. Nevertheless, they are essential if students are to learn successfully.

The authentic communication tasks themselves are entry points to developing a professional identity. For these students, the outcome of their BE Laboratory experience was not necessarily the mastery of specific tasks—an editorial rebuttal, a business plan, or a laboratory report. Rather, they were exposed to the range of discursive activities that professionals engage in and the ways that professionals attempt to make meaning by writing and speaking. Kuldell recognized and hoped to accomplish these more abstract outcomes: "I'm always aware that the details of what I'm teaching them are going to be lost in no time at all, and really that the best case scenario would be to have a more lasting impact on something that they will draw from, not in the detail of it, but in the sort of confidence that they have to, something truly foundational that rings true with them that they can develop a response that makes sense to them, and articulate it."

For students writing and speaking as professionals in BE Laboratory, these more abstract outcomes might be difficult to imagine as they struggle to draw a distinction between genetic engineering and synthetic biology, craft a business plan for the Registry of Standard Biological Parts, or analyze and communicate the results of their laboratory experiments. But these outcomes drive curriculum and instruction, and together they result in powerful learning experiences for these students.

3 Carving Out a Research Niche

The development of a scientific identity is not complete at graduation. Students continue to develop identities as professionals as they move into the workplace or graduate school (Winsor 2003; Berkenkotter, Huckin, and Ackerman 1991). The case studies in this chapter focus on a student's continued development of professional identity in graduate school. We examine how grant writing brings forth certain tensions as graduate students learn to define and motivate a research agenda. We also explore the pivotal role of a graduate student's mentor in this process. Specifically, we address these questions:

- How do students learn to motivate their scientific ideas in an organized fashion that appears significant to other scientists?
- What is the role of a student's mentor in learning to stake out a research territory?
- How do authentic writing activities like grant writing contribute to student learning?

Grants are a textualization of professional research identity. In grants, scientists stake out their research niche by claiming not only what turf they currently occupy but also the territory they seek to acquire. They lay claim to that desired new territory through acquisition of the research funds that allows them to generate data in that area (and subsequently generate research articles based on findings). Grants are thus a powerful assertion of individual identity for a professional scientist. As Gross points out, "Although the general progress of scientific knowledge relies heavily on the relative subordination of individual efforts to communal goals, the career progress of scientists depends solely on recognition of their individual efforts at discovery" (2006, p. 165). Latour and Woolgar explain in *Laboratory Life: The Social Construction of Scientific Knowledge* (1986) that the research profile of a scientist goes along with other forms of cultural capital, such as academic position, funding, and lab space. Professional scientists use that capital to gain credibility (and ultimately funding) in the scientific community. Scientists who "spend" their credits and credibility wisely are more likely to have successful careers. Researchers who squander their capital end up not conducting new research, hiring new investigators, or generating research funding. Once a scientist ends up in such a situation, "his grant money would be stopped, and, save for any tenured position or niche he has previously established for himself, he would be wiped out of the game" (Latour and Woolgar 1986, p. 230).

In addition to individual professional identity, the economics of scientific research are ever present in the professional lives of scientists. The ability to bring in research funding is central in developing a viable professional career, and grants are the primary written form by which scientists bring in research dollars. Whereas scientists must be able to provide motivation for their research agenda in research articles (Swales 1990), abstracts (Hyland 2004), research talks (Swales 2004), and poster presentations (MacIntosh-Murray 2007), grant proposals require them to be masters of persuasion. Grants are explicitly persuasive in their purpose, and although they may not carry the same kind of public prestige that a research article may provide, they show other researchers that a scientist has the ability to raise capital for a scientific enterprise. Thus, for students who wish to have their own lab one day, grant writing is an essential communication skill that must be mastered.

Grant writing may be considered a high-stakes form of writing. As former National Institutes of Health Director Elias A. Zerhouni explained in *Science*, there was a 70 percent increase in grant applications from 1998 to 2006: "In 1998, NIH received 24,151 applications for new and competing research project grants.... NIH expects to receive over 46,000 in 2006 and over 49,000 in 2007. (2006, p. 1088). Such intense competition in NIH grant funding means that only a small percentage of grants are funded, typically less than 20 percent (National Institutes of Health 2008b).

Grants are also an assertion of consensus among members of the scientific community. The peer review process used in grants certifies that the "candidate knowledge" proposed by the grant writers appears valid (Gross 2006, p. 99). Gross (2006) also contends that the knowledge certified through peer review is based on intersubjective argument. In other words, peer review opens scientific knowledge not to public scrutiny but to a representative sample of the scientific community. In order to remove some sense of arbitrariness from the peer review process, grants are scored against a set of general criteria. NIH grants are typically scored on five criteria—significance, approach, innovation, investigators, and environment—"to judge the likelihood that the proposed research will have a substantial impact" (National Institutes of Health 2004). These criteria draw attention to the persuasive features of the document, that is, whether the author has convinced reviewers that the research is important, timely, and likely to extend scientific knowledge (Inouye and Fiellin 2005). Rhetorically, researchers must show that they can (1) create, define, and refine the research project, (2) establish that the research will provide a significant contribution to the field, (3) situate the research within the literature of the field while distinguishing that research from work that has already been completed, (4) convey the unique scientific merit of the research while also maintaining the feasibility of the proposed plan, and (5) show that he or she already has the knowledge (or access to that knowledge) and resources available to accomplish the proposed research before being funded.

In a review of grant proposals, Myers found two of these features were particularly important: "How researchers situated themselves in relationship to the academic community—that is his/her status—as represented by the home institution, publications, previous funding, claims about the relevance of previous work, etc. [and] how well the researcher situated the planned work within the previous research in the field" (as quoted in Connor and Upton 1996). Myers (1990) showed how two biologists adapted these specific tactics in revising their proposals. These researchers tended to overstate their claims in their initial drafts, but their subsequent drafts asserted their authority and previous accomplishments with more force. Both used citations to advance their knowledge of the field or their own ethos by citing their own publications. Finally, both used specific terminology to indicate their "place in the community" (Myers, 1990, p. 51).

At MIT, the lore of grant writing is strong among graduate students, but many of them convey a sense that the grant writing process is too overwhelming for them to participate in early in their graduate careers. Nevertheless, faculty in the department have a strong desire to train graduate students in grant writing. And as the case studies show, students do learn from the experience even when their attempts are not so successful.

For graduate students, learning how to define and situate their own research within the previous research in the field is a key competency and a key assertion of identity. In addition, understanding how to sell a novel grant idea to fellow scientists requires an understanding of current research and what ideas are likely to appeal to scientific reviewers (Myers 1990). Graduate students, however, often find their knowledge of research in a field lacking when they attempt to write a proposal. MIT faculty member Sangeeta Bhatia explained:

> One of the most difficult things is to actually feel like you can have a new idea when you are a student. Because you walk into a field that's established . . . and you see the published papers out there. You feel, as a student, like everything's already been done. The literature is not written to

identify necessarily what *hasn't* been done, and where the important questions are. [It's difficult for you to] formulate the big picture of how to put it altogether, or how you're filling an important need, or why it will be impactful.... The second challenge is that in order to formulate a good question, you have to dig very deep into the literature to identify the goals. They sit down to write this set of specific aims, and they realize, oh my God, I really need to read 50 journal articles before I can even come back to this.

Graduate students are also unlikely to understand what ideas appeal to scientific reviewers. Although graduate students may have worked in labs during their undergraduate years or even coauthored articles with mentors, their sense of professional audiences is often limited. In order to gain an understanding of scientific audiences, students must engage in "authentic language practices"—repeated interactions with scientific audiences, using the professional genres of the practice. In her study of a physics professor, a graduate student, and a post-doc, Blakeslee (2001) found that such interactions with real audiences were paramount in gaining acceptance for their ideas. She also found that "distant guesses about audiences based on uniformity or broad-stroke characterizations aimed at manipulation and domination rather than cooperative interaction do not work very well ... real functional knowledge of audience comes only over time by entering into some community of practice" (2001, 11).

On a textual level, NIH grants pose another challenge for graduate students: the highly standardized format of the grant proposal limits the kinds of rhetorical appeals that a writer may use. However, while the highly standardized format of grants is limiting, writers can use a number of creative strategies within that highly structured format to make arguments for their proposed research. For example, the background and significance section of a grant appears deceptively like a literature review to novice readers. Yet, the background and significance section of a grant is not merely descriptive: it must contextualize why the particular approach being proposed is important in relation to other work in the field. The rhetorical challenge for grant writers is to review the literature in the field while also making clear the niche that their own research fills and that they are qualified to undertake that work. For example, a writer might take claim for his published work by citing himself or by using phrases such as, "we have previously shown ..." Writers may also choose to directly name certain prominent researchers in the field with whom their work aligns but not specifically name researchers whose work contradicts their own (instead preferring to lump them anonymously using phrases such as, "As others have suggested ...").

Grant writing poses a final challenge for graduate students. Because graduate students have little status as researchers, they must rely on their mentor's credibility in meeting the review criteria regarding the expertise of the researcher. In other words, while pursuing their own research agenda, students must work in the shadow of their mentors. In a year-long qualitative study of thirty-five graduate students in an NIH grant writing course, Ding (2008) found that graduate students who worked on grants relevant to their lab's projects had their work "incorporated into their advisor's larger projects" (p. 33). Thus, one of the most difficult challenges for science graduate students is performing the dutiful responsibilities of the apprentice scientist while staking out their own research territory. Graduate students find themselves trying to develop their own identities as researchers who specialize in the particular "study of" while maintaining a good working relationship with their mentors. Some graduate students simply perform the duties that their mentors give them. Others strike out on their own. For these students, the success or failure of their research often depends on whether their mentor decides that the work bears interest to the lab.

Learning How to Persuade Grant Reviewers in Frontiers in (Bio)medical Engineering and Physics

Frontiers in (Bio)medical Engineering and Physics (from here on, called Frontiers) is a small graduate class in the The Harvard-MIT Division of Health Sciences and Technology (HST). HST class was established as one of the first biomedical engineering graduate programs in the United States. The current curriculum reflects the varied research in contemporary biomedical engineering, ranging from imaging and medical physics to bioastronautics. One particular interesting feature of HST is that it allows students to have both the bench experience of a researcher and the bedside experience of a clinician. This dual identity of the program attracts graduate students who want a traditional career in research, as well as students who want to go to industry or medical school (Wilkerson and Abelmann 1993).

Sangeeta Bhatia, an HST faculty member, teaches the Frontiers course, which she initiated in 2003, and all of the students in the course are HST graduate students. One of the goals of Frontiers is to challenge graduate students to engage in the communication work of "real scientists," in particular, grant writing. The HST program also hoped that graduate students who were taught early in their careers about proposal writing would be able to use these skills while writing their thesis proposals and later translate those skills into NIH and other grant writing activities during their professional careers. In 2006 the HST program made Frontiers a required course for students concentrating in its medical engineering and medical physics specialization, and as of 2008, all HST graduate students were required to write thesis proposals according to NIH format.

Bhatia considered grant writing to be an ideal communication activity to teach graduate students to write about their own research. As she explained in an interview, grant writing teaches students "how to break [a project] into pieces that go with each other, how to define a beginning and an end of something, and how to communicate that in a way that is compelling to people who may be outside of [the lab]." Such tasks go beyond the tasks typically required of graduate students. Bhatia explained: "[The grant writing experience] may be the first time that they do not just read what they have been told to read for homework but to actually think in an open-ended way about what's interesting, what's important, what's unsolved, how to package it, how to communicate it, and then how to dig down into the details, so that you can write a compelling document."

Bhatia chose to use NIH R01 grants, the most prestigious of NIH grants, as the model for student writing in Frontiers. NIH R01 grants are investigator-initiated grants "related to the stated program interests of one or more of the NIH Institutes and Centers based on descriptions of their programs" and may be thought of as the gold standard of grant writing in the biomedical sciences (National Institutes of Health 2008b).

For students in Frontiers, this grant-writing experience may also be the first time that they engage in authentic peer review activities. In the case of Frontiers, authentic peer review means modeling the NIH study section reviews. Study sections are a form of team peer review used to evaluate the quality of grant proposals. Study sections at the NIH may include up to twenty-five members for the review of R01 grants. Although every member of the group reads the grants assigned to that study section, only two reviewers—a primary reviewer and a secondary reviewer—critique the entire grant in detail. The job of the primary and secondary reviewers is to address the strengths and weaknesses of the grants according to the NIH criteria. They begin the evaluation process by offering their scores before the other members of the study section assign their scores for the grant:

> Once the primary, the secondary, and the reader have delivered their opinions of a particular application, there is general discussion, and then the members (16 to 20 of them in the average permanent, chartered study section) assign priority scores between 1.0 and 5.0. Lower is better, with scores of 1.0 to 1.5 meaning "outstanding," 1.5 to 2.0 meaning "excellent," 2.0 to 2.5 meaning "very good," 2.5 to 3.5 meaning "good," and 3.5 to 5.0 meaning "acceptable." Once everyone has voted, the scores are averaged and multiplied by 100 to give a final score between 100 and 500. (Finn 1995)

Rather than designing the course solely around grant writing, however, Bhatia designed Frontiers to do two other things: provide technical presentations by other HST faculty and give first-year graduate students in the program a common core experience. In spring 2008, the semester in which this research was conducted, a different HST faculty member was invited to speak about his or her research at each class session. Five grant writing workshops punctuated the faculty lectures. According to the syllabus, the goals of the workshops were "to teach students the elements of successful proposals as well as familiarize students with the larger professional contexts in which scientific research proposals are written." Situating grant writing within the larger professional contexts in which proposals are written meant that students would learn not only the structure of scientific grants but also how scientists approach reading grants and the grant review process.

Each grant workshop focused on one section of an NIH grant: aims, background and significance, preliminary data, and research approach. In a final workshop, students completed a mock study section modeled after NIH study section reviews. The ordering of the workshops was not arbitrary; Bhatia wanted students starting "at the highest level." For example, with the aims workshop, she conveyed: "You have to start with how the project is going to be constructed. I'm of the belief that if you have the content designed, and the structure defined, that the writing will follow."

Bhatia's awareness of the challenges posed by the class activities meant that three objectives were repeated in the workshops: (1) breaking down a project into a series of steps through the aims; (2) organizing the rest of the grant, such as the background and significance section, around those aims; and (3) using rhetorical strategies throughout the grant to remind, educate, and persuade readers of the grant's merit.

In designing the workshops, Bhatia worked with a communication instructor to craft a specific set of activities for each of the workshops. For example, the aims workshop is described on the syllabus as follows:

Workshop #1: Defining a research topic: In this workshop, we will read specific aims of recent successfully-funded research proposals. We shall introduce the criteria by which such proposals are evaluated. By analyzing how such aims are written and how they are evaluated, you will learn how to define a research topic. Some of the questions that we will address in this workshop include: How do you define a research topic in the context of a field? How do you break down a research project into smaller components that build on one another (i.e., specific aims)? How much work should be encompassed in an aim? What's a risky aim?

As a final component of this workshop, we will then work from the specific aims of the sample proposals back out to a laboratory profile to show students how the research and thesis proposals generated within labs reflect the research focus and expertise of that laboratory.

At each workshop, Bhatia led students through an analysis of a sample grant. The decidedly rhetorical aspect of her commentary was one of the important features of the class workshops. Ai, the teaching assistant who was also an alumnus of the class and had worked extensively with Bhatia on grants from her lab, noted that this approach offered students "her perspective as an established PI [principal investigator]—how she sees things, like, 'that kind of significance to me as a reviewer means this, as a writer means this, as a scientist means this.'"

Bhatia's insider perspective on how reviewers approach grants provided students tips on how to make each part of a grant more persuasive and, ultimately, more likely to be funded. For example, in the workshop on specific aims, she explained that grants typically have three aims, with the first aim being the "safest," meaning that most of the research should already be underway for this aim, and the third aim should be the "riskiest," meaning that the least amount of research should be done on that aim. She also explained that while the aims should build on each other in a "crescendo," they should also be independent enough from each other that if one aim fails, the project can continue. Ultimately the rest of the grant follows from the aims, so if the aims are not clear and well organized, the reader is unlikely to view the rest of the grant favorably.

At the end of each workshop, the fifteen students in the course were to write that section of the grant for their own research. This process continued for each workshop and culminated in students writing a complete grant by the end of the semester. Students were given feedback with each submission from the course TA. Bhatia reviewed the students' aims until she felt their projects were focused and feasible, and she reviewed the final grant submissions at the end of the semester.

At the end of the semester, students participated in a mock NIH study section. Each student was assigned as a primary and secondary reviewer on two grants. Modeling the roles of a NIH reviewer, Bhatia showed the students how to score and respond to a grant, giving "honest feedback but couching it in constructive terms." Bhatia's desire here was that students engage in an authentic form of peer review and thus remove "the teacher" from some of the evaluation process.

Although the goal was to provide an authentic peer review process for students, the NIH guidelines had to be truncated to accommodate the student nature of the grants. Students were not asked to evaluate the expertise of the investigator or the environment in which the work was being conducted. They had one week to score and respond to peer grants, using the remaining three NIH criteria:

1. Significance. Does this study address an important problem? If the aims of the application are achieved, how will scientific knowledge or clinical practice be advanced? What will be the effect of these studies on the concepts, methods, technologies, treatments, services, or preventative interventions that drive this field?

2. Approach. Are the conceptual or clinical framework, design, methods, and analyses adequately developed, well integrated, well reasoned, and appropriate to the aims of the project? Does the applicant acknowledge potential problem areas and consider alternative tactics?

3. Innovation. Is the project original and innovative? For example: Does the project challenge existing paradigms or clinical practice; address an innovative hypothesis or critical barrier to progress in the field? Does the project develop or employ novel concepts, approaches, methodologies, tools, or technologies for this area? (National Institutes of Health 2004)

In the final workshop, students were divided into study sections. Each primary and secondary reviewer began the session by offering a score for the grant. The primary reviewer then presented the grant to the rest of the study session reviewers, offering his or her assessment of the significance, approach, and innovation of the proposed research. The secondary reviewer offered an additional assessment of the grant's merits and weaknesses. Next, other reviewers could ask questions about the content of the grant. Following discussion, the reviewers offered their final scores of the grant. If scores deviated by more than a few points, deliberation continued until reviewers were in closer agreement.

Following the study section, reviewers' comments were returned to the grant authors so that the authors could "respond" to the comments in a letter to reviewers. They could also revise their grants to reflect reviewers' concerns. Seventy percent of students' final course grades was based on timely submission of drafts (30 percent), quality of the final, revised grant (25 percent), and quality of study section reviews (15 percent).

Learning How to Sell Science

The case studies in this chapter focus on three students who enrolled in Frontiers in spring 2008. These students are representative of the others in the class in that their struggles and successes with grant writing were similar to those of their classmates. In addition to these case studies, we also surveyed the Frontiers students at the beginning and end of the semester.

Our initial survey asked Frontiers students about the purpose of each section of a grant as well as the NIH review process. The pretest survey results showed that students had a good sense of the purpose of each section of the grant but lacked a sense of the scope of an NIH R01 grant. For example, some were unsure if the grant should describe expected findings, thought that the same amount of preliminary data should be provided for each proposed aim, and thought that a single grant could typically have up to six measurable aims. Figure 3.1 shows pretest and posttest results from four of the

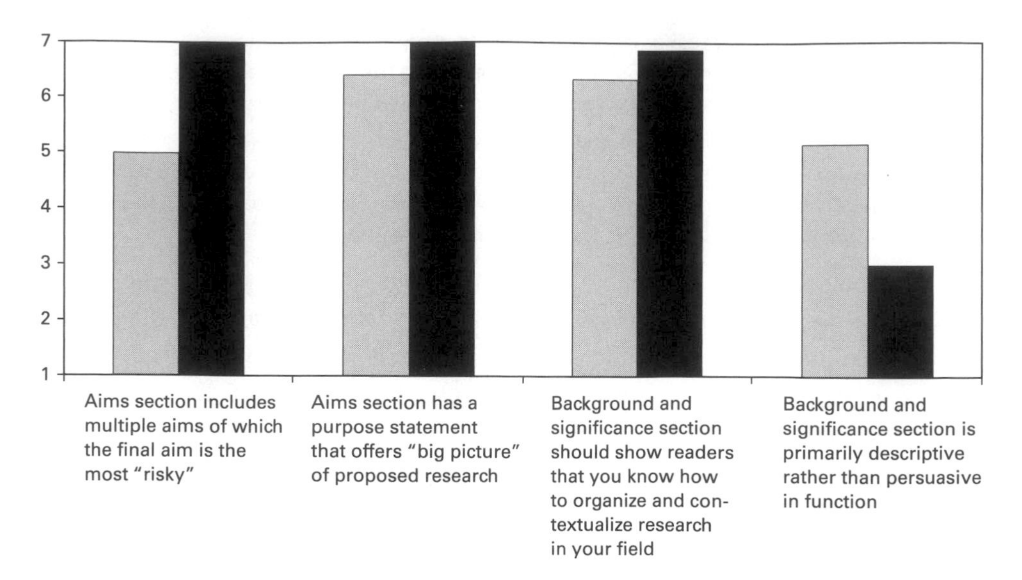

Figure 3.1
Student pre- and postsurvey answers. The gray bars are the pretest data, and the black bars are posttest data. Surveys were answered on a 1 to 7 Likert scale, with 1 being "strongly disagree" and 7 being "strongly agree."

survey questions. By the end of the class, students had come to understand that the final aim should be the "riskiest" of aims. They also came to a fuller understanding of how to contextualize their research and the literature of the field. Finally, by the end of the semester, students had come to understand that the background and significance section of the grant should be persuasive, not descriptive, in nature.

Students also lacked a definitive sense of NIH review processes. They were unsure how to answer questions about the NIH review process: for example, "All NIH grant submissions are given a written review" and "You may appeal your reviewer's scores" (For a complete list of questions, see appendix B.) The findings confirmed that students had little experience with NIH process. Their experiences were captured by Alyssa's explanation: "I have no idea how NIH actually works because everything I knew was from my PI, who [just] told me what I needed to write. I didn't know who was reading it, and how it's read in order to [score] it."

In the following case studies, we look at three students who had quite different experiences in Frontiers. Joe came to the class with an idea for a research project but was not certain about the direction of that research. Over the course of the semester, he not only developed his project but also persuaded his mentor to take a serious interest in that project. Alyssa came to the course with a decidedly graduate student perspective: she intended on using the proposal produced in the class as her thesis proposal. And although she accomplished that goal, Alyssa's mentor still had not read her proposal by the end of the semester. Like Alyssa, Park came to Frontiers with a project: continue research from his master's degree work. Yet, he struggled to capture his mentor's interest in this project, and at the end of the semester, he had changed his research focus to better accommodate his mentor's interests. For Park, however, Frontiers was not a failure. Like Alyssa, he was an international student who had little knowledge of U.S. grant funding. Frontiers was an opportunity for him to learn about U.S. grant funding for future collaborations as well as an opportunity to work on his second-language writing skills. These students' stories highlight the central role that mentors play in the research lives of graduate students. For graduate students, convincing one's mentor that one's work is important is just as important as convincing one's peers. Their stories also highlight how tensions over what is persuasive and what is not arise in the evaluation of peer work.

Together, the case study data and survey results point to three major findings regarding how students learn to define and motivate a research identity while negotiating that identity within the context of their relationship with their mentors:

- Learning to define a project and organize the literature of the field was an iterative process for students.
- While the grant writing activities took place within the space of a classroom, the influence of mentors was a mediating force on student learning.
- Engaging in authentic forms of peer review helped students better learn about the persuasive strategies used in grant writing, but it also brought forth interpersonal tensions in peer evaluation.

Case Study 1: Joe—Learning to Sell Science

Joe saw Frontiers as a good opportunity to get back into research that he had started in fall 2006 but had not had time to work on since then because of medical school classes. Joe was not committed to a specific approach for that research and wanted to use Frontiers to develop and focus his research approach: "to kind of mature my hypotheses of what I'd really like to do."

Joe explained his project as "a branch of a current focus in our lab, to look at tissue-engineered cellular constructs in order to study the disease process and maybe to suggest new therapies for [cancer]." Joe explained that recent research in angiogenesis has shown that tumors need specific types of blood vessels that increase the tissue pressure by the tumor and thus force migrating tumor cells to other regions and enable hypoxia, which is important in the development of virulent tumors. Research has also shown that noncancer cells play important roles in cancer because they are "recruited" by tumors: "tumors recruit kind of a team to help them grow." Joe's work focused on how "endothelial cells recruited within vessels themselves play a role in encouraging cancer."

In the first draft of his grant after the aims workshop, Joe began broadly with the clinical motivation for his research. After briefly mentioning angiogenesis, Joe then discussed the lab's previous work in the area of endothelial cell research, subsuming his own research contributions with those of the lab. In describing the lab's previous research on endothelial cells, he made a connection to anti-angiogenesis: "Our work ... on tissue repair leads us to a potential third mechanism, namely that the vasculature is a source of uniquely potent regulatory cells, those of the endothelium." Joe's next strategy was then to use a series of hypothesis statements to suggest how endothelial cells might be linked to formation of blood vessels. In presenting his aims, he chose to add one or two short explanatory sentences after each aim, such as shown in his first aim:

> **SPECIFIC AIM 1: To investigate how TEEC [tissue-engineered endothelial cells] regulate the phenotype of well-differentiated and poorly-differentiated cancer cells in vitro.** Many carcinomas are continually dependent on stromal cells for proliferation. Therefore we plan to use TEEC constructs as in vitro regulatory tissues. We will study the proliferation of well-differentiated and poorly-differentiated cancer cells exposed to TEEC constructs and we will characterize the proliferative, invasive, and metastatic markers of these cancer cells pre- and post-exposure.

Joe described the organization of his aims as "practical," beginning with in vitro projects that the group had already started: "In medicine, biology and science, if what you're studying has biological relevance, you want to show that it works [in vitro]. You do not want to just start directly on a mouse or a rat."

Ai, the teaching assistant, had concerns about Joe's draft, as she explained in her feedback to him:

> This is a good start and an interesting project, but overall goal and hypothesis are very scattered. First, considering overall goal/big picture, 1st paragraph suggests your research will be on anti-angiogenesis, but 2nd paragraph suggests that you are interested in elucidating fundamental mechanisms of endothelial cells in tumor biology. Should focus on one and convince reader why it is important.... You also want to organize your aims in parallel with your hypothesis, in a logical progression (where aims build upon each other's findings).

Ai was concerned that Joe's proposal was not focused and that he seemed to be working on anti-angiogenesis and tissue engineering without explicitly showing how the two fields were related. She was also concerned about the link between Joe's hypothesis and aims: that Joe had multiple hypotheses that were not linked to specific aims. One possible strategy in writing the aims section of the grant, as Bhatia explained in the aims workshop, is to have a central hypothesis and then a supporting hypothesis for each specific aim. She had also stressed the importance of making connections across various fields clear when working in an interdisciplinary field like biomedical engineering.

Joe was receptive to Ai's feedback. In describing his strategy for revising his aims, he was particularly focused on making the link to the hypothesis clearer: "I'm going to try to make everything consistent with the hypotheses, the main hypothesis as a kind of corollary, and I'll try to tie better with the aims."

Joe's revised aims show that he had taken Ai's advice, more clearly linking angiogenesis and tissue engineering. He also foregrounded a singular hypothesis before his specific aims and added an explanation after each aim to further detail the proposed research for that aim (box 3.1).

In discussing his revised aims section, Joe described this particular revision as "crucial" because "it's important to gather the reader in the first page": "I've never really had to try to persuade people like this before. It's much more difficult than I thought—to present your ideas and try to sell the ideas on just one page, basically a paragraph.... I guess it's much more difficult than I thought, but it's also a very good learning experience."

In the description of his revision, Joe adopted the discourse of "selling science." He had come to sense the exigency in gaining reviewers' trust and interest in the first page of the grant. Unlike a research article where readers may skim the introduction of the article, the first page of the grant is considered the most important page of the document. In Joe's comments, we see him coming to an awareness of this different reading style from readers and how that changes the rhetorical context. Other students, like Alyssa in the next case study, would struggle with that sensibility. The test for Joe would come at the end of the semester: Was he persuasive enough to get his grant funded?

Joe's rhetorical task in writing the background section to his grant was to introduce a new direction of research in the study of tumor formation. He had a difficult challenge in that new theories about cancer stem cells had recently become popular, so he needed to address how those theories fit (or did not fit) with his own research agenda. In terms of motivating his research for readers, Joe employed a number of rhetorical strategies, such as citing a seminal paper in his field by the author's name and addressing anticipated rebuttals of his approach. He had taken Bhatia's advice on how to

Box 3.1

Joe's revised aims for his grant, "Defining the Regulatory Roles of Endothelial Cells in Cancer Angiogenesis"

The development and spread of many cancers depends on their local blood supplies. The sprouting of new blood vessels from existing blood vessels, angiogenesis, has therefore been a focus of cancer research. Blood vessels can provide perfusing pressures and nutrients that may be rate-limiting for tumor growth (1–3), or by virtue of pathologic permeability increase local pressure, alter intratumoral hypoxia, and facilitate outward migration of cancer cells (4–6). Our work defining the paracrine effects of endothelial cells (EC) on tissue repair leads us to propose a *third* role for tumor vasculature. As in all vessels, tumor vessels are comprised of EC, supporting matrix, and occasional supporting cells at the perimeter. EC serve as vascular epithelium with potent sensory, secretory, and tissue-regulatory attributes. Fine local control arises as each of these attributes is responsive to subtle focal stresses with profound changes in EC genotype and phenotype. We recently showed that the epithelium of complex injured tissues is preserved by the tissue vasculature EC (7). These findings can be extended to cancer, which seemingly are sensitive to vascular control and are usually epithelial in origin. We hypothesize that, in addition to effects related to their perfusion function, vessels serve as scaffolds that retain regulatory EC in the vicinity of the tumors....

Specific Aim 1: To define how TEEC regulate the phenotype of well-differentiated and poorly-differentiated cancer cells in vitro. We will determine how TEEC constructs regulate the phenotype of well-differentiated (NCI-H441 lung adenocarcinoma, SK-LMS-1 leiomyosarcoma) and poorly-differentiated (PUB/N lung carcinoma, SK-UT-1 leiomyosarcoma) cancer cells *in vitro.* In particular we will determine whether the control TEEC elicit over hyperplastic vascular smooth muscle cells (SMC) or bronchial epithelial cells can be recapitulated in *neoplastic* cells. We will characterize the effects of TEEC conditioned medium on the proliferative (cell number, MTS assay, ^{3}H-thymidine incorporation) and invasive (transwell migration assay) phenotype of cancer cells and on their expression of cancer biomarkers (p53, Rb, HIF, and others) and compare these effects to those observed in smooth muscle cells. In turn we will then molecularly modify EC—via siRNA knockdown and forced overexpression—so as to regulate their production of factors critical to control SMC hyperplasia (e.g. HSPG, TGF-b) and determine if the TEEC's cancer-regulatory phenotype is similarly sensitive to such factors.

structure the background and significance section persuasively. For example, during the background and significance workshop, Bhatia had stressed that this section of the grant shows scientific reviewers that a writer knows how to synthesize the literature in the field, how to organize the literature across several fields (because biomedical research is interdisciplinary), and how to choose important literature in the field to cite. The background section should always include pointers to the researcher's project, with constant, subtle reminders back to the overall focus of the grant. Joe had adopted these strategies in his synthesis of the research literature, showing not just the context for his work but also the significance of his work.

In writing his draft, Joe had picked up on several key points from the background and significance workshop session. He started his background and significance section with the clinical motivation for his work and then quickly focused down on his area of research. He made that transition by discussing "one of the hallmark review articles," which happened to be written by one of his mentor's collaborators. Joe anticipated a potential rebuttal from his readers and added a section on cancer stem cells "just to make the reader know that I acknowledge that these exist." Following a discussion of cancer stem cells, Joe explained specific interactions between cancer cells and their supporting environment to show what has "been studied and validated and shown to be very important." The final section of the background section was where Joe advanced his own ideas about the relationship between endothelial cells and tumor formation and inserted (again) his primary hypothesis. Because his mentor had already worked in this area, Joe could situate his ideas within the safety of the lab's existing research: "Our laboratory and others have shown that bone marrow EC in hematologic malignancies have an activated phenotype and that the activation of quiescent EC is important for angiogenic neovascularization and cancer virulence." His own hypothesis was couched within the language of the lab "we" identity: "Our hypothesis is that, in addition to the perfusion-related mechanisms of Folkman and Jain, the tumor vessels [endothelial cells] themselves are potent regulators of tumor virulence, i.e. some tumors are caused by loss of endothelial or epithelial (since endothelium is a type of epithelium) growth control." The last paragraph of the background section Joe described as a way of "summing up in a compact way."

In describing how he crafted and revised the various sections of his grant, Joe described the hands-on approach that his mentor, Eli, took in helping him. Initially Eli was not so keen about Joe's grant writing assignments but began to take an interest in Joe's writing after the feedback on Joe's aims section. Now, before each submission, Joe and Eli sat down and went over Joe's draft: "On the average, [Eli] and I went over it face-to-face for 20–30 minutes. The criticism he had was usually less about specific wording and more about the big picture, saying this organization here or maybe you should rearrange section X." After several iterations, the feedback from Eli became a pivotal aspect of Joe's relationship with his mentor—a relationship that would have both short-term and long-term positive consequences. In the short term, Eli had extensive input on the research plan section of Joe's grant "because he had a lot more experience in trying to lay a logical map of what you were going to do." In the long term, Eli's involvement would mean that Joe's grant would have a life beyond Frontiers.

Perhaps in part because his relationship with his mentor was so collaborative, Joe took a "diplomatic" approach, as he described it, to his study section reviews. In describing how he responded to other grants, Joe said he found it "interesting . . . going through and seeing other people's thought processes and trying to think of how to improve that and also broaden your own references." Joe rationalized that other reviewers would be "fair" too and "suggest reasonable things to make [a grant] more persuasive or more clear."

It was in the study section process, then, that the concept of selling science took on meaning for Joe. In the study sections, reviewers present grants to other reviewers, and the skill with which they present another scientist's research has a strong influence on the score that the grant receives. Joe found it "discomforting" that members of the study sections could score grants "based on how hard the person vouches for you. Or how well the primary reviewer portrays, because they *buy into* your proposal. . . . Potentially only [a few] people read your grant in real life and that dictates whether or not you get funded."

By the end of the semester, Eli's collaboration with Joe on his grant paid off with an unexpected outcome: Eli suggested that Joe add more to his grant and submit it to the NIH. This outcome was important for Joe because it showed that his mentor believed in the quality of his research. Besides, as Joe explained, it "never hurts to try get more money." At the time this book was published, Joe and Eli were still waiting to hear from the NIH reviewers about Joe's grant. Joe said in an e-mail, "As with any R01 the odds are slim, but hopefully (if we don't get it the first time) we can build on positive comments and eliminate problem areas to get it upon resubmission."

Joe's case study illustrates several pivotal moments where graduate students come to see science as a human enterprise in which persuasion shapes how science itself proceeds. Joe's relationship with Eli shows how dramatically a mentor can shape the future of a graduate student's research career. Through subsequent iterations of Joe's proposal, his mentor, Eli, became more convinced of its merit. However, Joe's proposal was not as convincing to his peers. Although the grant scored highly, it would not have received support if his peer reviewers' scores were actual NIH scores. In the current NIH climate, approximately only the top 20 percent of the highest scoring grants receive funding.

Case Study 2: Alyssa—Striking Out on Her Own

Alyssa was the only student in Frontiers who self-identified as a geneticist, and she came to the class with a clear purpose in mind: to write her thesis proposal. Alyssa, however, was ultimately frustrated by the classroom aspects of the peer review and sought an authentic scientific readership for her grant. Her frustration stemmed, in part, from the fact that she approached the grant writing activity with a strong graduate student identity while seeking a professional reader as the audience for her work. Alyssa found "motivating" her research frustrating because of the seeming contradictions in the student audience in Frontiers and the intended audience she imagined of professional genetics researchers.

As Alyssa explained in her first interview, she came to the class with a good sense of her project and had already identified a mentor: "I've already had this figured out a long time ago, last year, and I talked about it with my PI [principal investigator]. All these ah-ha moments happened last year. So I pretty much already had it figured out what I wanted to do ... but now I actually got down to business to determine exactly what needs to be done, in what order, and what kind of experiments does it entail."

Alyssa's thesis project focused on the genetic components involved in understanding autism. As she explained, researchers know that autism is genetic but do not know the genetic underlying mechanisms of the disease. In the near future, researchers hope to better understand the genetics of autism in order to find better diagnostics to diagnose the disorder and develop better interventions for managing it. In her research, Alyssa specifically focused on small RNA development because small RNA molecules have been found to be a key regulator in brain development. Autism is likely caused not by a single genetic variation but by a complex series of variations. In her proposal, Alyssa characterized this purpose as to "discover sequence variation and subtle differential expression of small RNAs" in autism.

As Alyssa explained, the project reflected her own original hypothesis (not her mentor's), and she was "really eager to test it." Being able to design her own set of experiments around her own hypothesis was an accomplishment for a graduate student. For the past two years, she had worked for her mentor on his projects, and her thesis marked an opportunity for her to step out from his shadow and take on her own identity as an independent researcher. Her thesis proposal was a step toward that goal because it laid out the experiments she would conduct to gather the preliminary data to advance her research.

While Alyssa had a detailed sense of what experiments to undertake and how to stage or organize those experiments, an underlying issue in her proposal was that the design of her project had to be "efficient" because she did not have a grant budget of her own to carry out the experiments: "I want to be able to test my hypothesis with a minimal amount of experiment.... Because right now, of course, nobody's going to give me money for doing some grandiose experiment that costs $50,000. So I have to come up with an efficient way to test this hypothesis." As Alyssa saw it, trying to persuade a fictitious NIH reviewer to fund her research was pointless; she was a graduate student who needed to convince her PI that she could do her experiments cheaply.

Alyssa's sense of her mentors as audience for her grant was reflected in her first draft. Like Joe, Alyssa began with a broad clinical motivation for her work, calling autism "one of the most devastating diseases of childhood." She then turned to the genetic basis for autism and submerged her discourse in the language of a specialist in justifying her approach: "In this proposal we use massively parallel single molecule sequencing to discover sequence variation and subtle differential expression of small RNAs and their targets in autistic and control brains." Alyssa used a similar discursive strategy for her aims, relying heavily on the specialized language of genetics research, as shown in the following example from her grant:

Specific aim 1: robust sequencing and quantification of small RNAs in autistic and normal brains. Hypothesis: microRNAs and snoRNAs orchestrate complex brain development (19). Both molecules are heavily involved in RNA editing (20,21), which is crucial for normal behavior (22–24). Alterations in these upstream regulatory mechanisms can account for the broad phenotypic spectrum and complex inheritance pattern observed in autism (since the regulators affect multiple transcripts, each subject to sequence variation of its own).

Size selected RNAs from normal and autistic cerebella will be sequenced on the 454 FLX system, following well-established protocols (25–28). We will focus on 5′ phosphorylated 19–28nt (microRNAs) and 80–95nt (HBII-52 snoRNAs) molecules. Data analysis will center on identifying sequence variation in autistic vs. control RNAs; identifying novel RNAs according to their folding characteristics; identifying the genomic positions of detected RNAs and their relationships with established autism linkage and association regions; assessing the differential expression of detected RNAs in autistic vs. normal brains; identifying A-to-I editing of small RNAs by a comparison to the genomic sequence; quantifying differential editing of small RNAs in autistic vs. normal brains; and inferring the function of differentially-expressed or differentially-edited RNAs and their correlation with the autistic phenotype by computationally identifying their protein-coding targets.

Each of Alyssa's aims reflected what she saw as a feasible outline for her research in the next several years while finishing her graduate work. Alyssa explained that the first two aims were "easy to carry out" and "narrow." The third aim built on the first two. Alyssa described this aim as "riskier" than the first two aims yet still feasible.

In discussing her aims, Alyssa picked up on some of the concepts that Bhatia had discussed in the aims workshop, namely that aims should build on each other but also should be independent. Some aims should be "safe" or "narrow," and others should be "riskier." When asked to define what was significant about her project, Alyssa was quick to note that only a reviewer who was familiar with research over the past year in her field would understand the significance of her approach. Studying autism has obvious significance, but only recently have researchers started investigating small RNA involvement in its development. She was also aware of potential criticisms of her research given the limited brain samples and that the sequencing techniques may show no differences between the two populations. She hoped that her cost-effective, manageable experiments, however, would make her project appear less risky to a reviewer, who might "give her a chance."

Ai, the teaching assistant, liked the structure of Alyssa's project but noted that the language needed to be more accessible to a nongeneticist: "Nicely written statement with well-defined plan and goals. Main suggestion is that it was challenging for a non-expert reader to make sense of a lot of jargon or to really be sure that goals are achievable.... This is the section of the grant where you should speak most generally to the widest audience, so be careful not to go over readers' heads." Ai also suggested that Alyssa add a stronger statement regarding the significance of her research, "which can adequately motivate why such in-depth understanding of autistic RNAs and their regulation at multiple levels is so critical to our understanding/treatment of the disease (and therefore worthy of investigation)."

Alyssa took Ai's suggestions, although she was somewhat reluctant to change the prose style. She added more details for the motivation of her research but did not make the language substantially more accessible for nonexpert readers. Alyssa saw professional genetics researchers as her only audience for her grant and argued that dense technical language was necessary to convince them that her research path was plausible (box 3.2).

In discussing her limited revisions to the aims section, Alyssa said, "I was thinking that the proposal is intended to serve as my thesis proposal, to be read by professional geneticists, and I did not care to change it just for the exercise of class. I wanted to have a finished product with your and Ai's feedback that suited my needs."

In writing the background and significance section to her grant, Alyssa's rhetorical task was to persuade her readers that new technologies could be used to study autism. Here, she spent more than one-third of the section discussing the history of autism research or, as she explained it, acknowledging the researchers whom she had envisioned as the reviewers for her project. In the next subsection on small RNAs in brain

Box 3.2
Alyssa's revised aims for her grant, "Small RNAs and editing in autistic brains"

Autism is a common neurodevelopmental disorder with a significant genetic component, characterized by a spectrum of social deficits and repetitive behaviors[1]. With a prevalence of 1 in 152 individuals[2] and lack of effective therapy, autism is one of the most devastating diseases of childhood. Understanding the genetic basis of autism is needed to improve diagnosis and to improving approaches to therapy. Twin and family studies provide substantial evidence that autism is amongst the most heritable complex disorders[3–5], yet numerous linkage and association scans[6–10], as well as direct sequencing of nearly 200 candidate genes[11–14], have had little success in determining the genetic basis of autism. As RNA-mediated post-transcriptional regulation gains recognition for its key role in normal brain development[15–19], it becomes a plausible candidate for the elusive underlying cause of autism. Our goal is to investigate, for the first time, the involvement of small RNAs and editing in autism. In this proposal we use massively parallel single molecule sequencing to discover sequence variation and subtle differential expression of small RNAs and their targets in autistic and control brains.

Specific aim 1: robust sequencing and quantification of small RNAs in autistic and normal brains. microRNAs and snoRNAs orchestrate complex brain development[20]. Both molecules are heavily involved in RNA editing[21,22], which is crucial for normal behavior[23–25]. *We hypothesize that alterations in these upstream regulatory mechanisms can account for the broad phenotypic spectrum and complex inheritance pattern observed in autism* (as the regulators affect multiple transcripts, each subject to sequence variation of its own).

Size selected RNAs from normal and autistic cerebella will be sequenced using the 454 technology, following well-established protocols[26–29]. We will focus on 5′ phosphorylated 19–28nt (microRNAs) and 80–95nt (HBII brain-specific snoRNAs) molecules. Data analysis will center on identifying sequence variation, differential expression and differential A-to-I editing of autistic vs. control RNAs. Sequence reads will be mapped to the human genome, directly counted and compared among ASD, control and the reference genome.

development and disease, she focused on research from the previous two years because autism researchers have gotten "a lot of money" from Congress to study the genetics of the disorder. The remainder of the background section focused on the area of research that interests Alyssa. As Alyssa explained, she detailed the molecules that she wanted to study to show the reader that knew "what [she] was talking about" and to convince him or her that this focus was "important." She then described three autism-related genes and explained that although researchers know these genes are related to autism, they do not know why. She concluded the background section by returning to the clinical need for her research: "because little kids are suffering." Alyssa saw the background section as establishing the clinical need ("why we need to study autism") and cost-effectiveness of her approach as well as reviewing what has been done in the field ("because we do not work in a vacuum") and to make the case that the previous two decades of research on autism have yielded "nothing," thus the field needs "an original way to study autism genetics." This last goal would allow her to discuss what was novel about her research—using mRNA sequencing to study disease—while also showing that she had an understanding of other techniques typically used in genetics research.

In discussing her rhetorical choices for the background section, Alyssa intended genetics researchers, such as her mentors, as her audience. She wanted to convince them that she had done her due diligence in reading the seminal figures in the field for the past two years. Again, the niche she carved out for herself was a graduate student niche, and throughout the grant, there was a sense of the graduate student identity at play—a graduate student who needed to convince her professional genomics mentors.

Yet that audience was not the audience for Alyssa's grant in Frontiers, and that disparity created a problem in her study section review. According to Alyssa, none of her reviewers sufficiently understood her material well enough to critique her grant. Worse, they seemed to put little effort into understanding her research. She wrote in a note to Bhatia with her final draft:

> I just have to note that I had to ignore a significant part of my reviewers' comments because they were not feasible (e.g. test findings in a model organism → NO AUTISTIC MICE). . . . Actually I was pretty disappointed that they didn't get the gist of my proposal: they couldn't even summarize my approach correctly. Of course at first I blamed myself, so I let some others read it and they got it pretty quickly. Also Ai has read my proposal throughout the semester and understood it perfectly with no genetics background (and suggested some really good remarks—thanks!). So I concluded that the reviewers had not spent too much time reading this. I understand that in a real study section the reviewers will also not spend much time on each individual grant but when we share the same field, it does not take much effort to see why a proposal is innovative and significant.

While Alyssa saw the problem as the reviewers' lack of knowledge and effort, the reviewers had suggested that part of the problem with Alyssa's grant was the writing: "The reviewers [said] that if I had written it a little better, they might have had an easier time understanding what I wanted to say." The source of the tension between the reviewers and Alyssa was an issue of audience: Was this class exercise authentic in that Alyssa should be writing to the audience she intended for her thesis proposal or the audience of student reviewers in Frontiers? Who was Alyssa supposed to be convincing that her research had merit?

Alyssa's frustration for her peer reviewers was compounded because she brought a different expectation for their behavior to the review process. Her sense of how to approach professional reviewers was based on "respect" for those reviewers, a value she had learned through her interactions with one of her mentors—a "hard core scientist"—who had taught her to "not insult the intelligence" of reviewers. This mentor's rhetorical approach to motivating his research for grant reviewers was to be "meticulous" in the description of his work, which Alyssa described as "persuade that it's important but on a very detailed level." She saw this mentor as being able to motivate his research without overtly selling his science to reviewers, which Alyssa saw in grant proposals that sounded too much like "popular science." In contrast to this mentor, Alyssa described her second mentor as a "businessman who comes to sell his product." Her "businessman" mentor was highly successful in obtaining grant funds, but Alyssa felt that the claims in his grants often went too far, and she felt uncomfortable making claims that she could not accomplish in the grant time frame: "I do not say that I'm going to cure autism.... Practically speaking I'm just going to perform mechanisms that will later help other researchers ..., but it's really a long way to go.... I prefer to be more down to earth and a little more humble about what you can promise that your grant can do, but ... if you make all these grandiose promises then you might be able to persuade a larger audience. But I do not feel comfortable with it."

In her revised grant, Alyssa saw herself trying to find "a middle ground" between the "business woman" and "hard core" scientist rhetorical stances. Yet her uneasiness with the overt selling of scientific research brought up another problem in her study section: the grant reviews she gave to other students. She was particularly concerned about one grant in which the writer seemed to have insufficient background knowledge to carry out the proposed research. The problem was compounded because of a conflict of interest between the writer and the primary reviewer—the primary reviewer, friends with the writer, gave his friend a good score on the grant. Again, although Alyssa was "annoyed" by the apparent conflict of interest, she returned to a professional identity as a "safe house" strategy in reviewing the grant. She explained why she gave the grant a poor score:

> You have to convince me that you can do it. If you make mistakes, [you cannot] demonstrate that you know the biological processes that you're trying to study. So I do not believe that you can actually study them if you didn't take the time to write about it in correct detail.... So my criticism was pretty constructive in saying maybe find out some more research, do not even suggest to do really sophisticated stuff without knowing the basics for doing those experiments.

The conflict with the other reviewer at her study section left Alyssa with a negative impression of the peer review process used in scoring NIH grants, "It's just a way to get more enemies," she commented. Yet beyond her negative experiences with the study section review, Alyssa also said that she left class with a better sense of what affects "an average reviewer," the importance of specific aims, "tricks" to get a reviewer's attention, and ways to "suck up intelligently."

In the end, Alyssa's proposal would have a second life as her thesis proposal, but she thought that it had few other possibilities, mainly because the lack of preliminary data. As she explained, "At the NIH, if you have no preliminary data, they do not even look at you." At the end of Frontiers, Alyssa gave her proposal to her mentors. She commented in our final interview, "They didn't read, but at least I gave it to them." But Alyssa's experiences in Frontiers were not without an outcome. At the end of the semester, she was helping write the experimental design section of an NIH grant for her lab. As she explained, she was given the task only because "nobody wants to do it, and they think that it's not important." Alyssa's story took a surprising turn when we talked to her eight months later. Her mentors had read her proposal and liked it so much that they encouraged Alyssa to submit a shorter version of her Frontiers grant to the Autism Foundation for funding. At the time this book was published, Alyssa's grant was still under review.

Alyssa's case, like Joe's, illustrates how dramatically mentors can influence graduate students' experiences and their view of science. Alyssa's struggles over "real science" versus the "businessman" approach to grant writing left her uneasy with her own reviews as well as her peers' work. Although Alyssa felt frustrated by the study section review process, she had learned an important lesson about the subjective nature of peer review. Her sense of responsibility as a reviewer was also an outcome of her interactions with her mentors and clearly reflected her emerging professional identity as a contributing member of the scientific community in which she was working.

Case Study 3: Park—Trying to Find an Audience for His Research

Like Joe and Alyssa, Park came to Frontiers with a research project already started. The project he used as the basis of his Frontiers grant proposal was a continuation of his master's research, which he had worked on for the past two years. Park was interested in using optics to study human disease; he chose to work with red blood cells (RBCs) because they are optically simple. Specifically, his research focused on quantitatively characterizing the membrane fluctuations of RBCs. Scientists have discovered that the RBCs of patients who suffer from malaria and other diseases become rigid. Researchers are unsure why this happens and seek a better understanding of the physiology of RBC membrane fluctuation. As Park explained, "Actually [the membrane fluctuation] phenomenon has been known for a long time, but here we propose a mechanical way to interpret the membrane fluctuation data."

Park was interested in grant funding so that he could develop better methods to "prove" his technique with clinical data. Although he was interested in learning about grant writing, he had little knowledge about the NIH funding process or generally how to structure a grant proposal. As he explained, "In our lab there are very few graduate students. Most research [and grants are] done by the post docs and the scientists."

In the scientific community, Park's previous research would be considered basic science research because it did not have an immediate clinical application. Thus, one of the challenges for Park in Frontiers was to relate his work more clearly to a clinical application.

Park faced two other challenges as well. First, his lab mentor was not very interested in Park's work on RBCs and wanted him to finish this research as quickly as possible to move onto "totally different stuff." A second challenge for Park was that unlike Joe and Alyssa, English was not Park's first language. Although his language skills were advanced and he had no problems producing the twenty-five-page single-spaced grant by the end of the class, Park's writing and speaking were sprinkled with common errors, such as missing articles and verb tense errors.

In the first draft of his aims section, Park did not start with a broad clinical application like Joe and Alyssa. Instead, he began by explaining the goal of his research: "to establish the integrative micro fluidic imaging technique to retrieve materials properties of single cell from membrane fluctuation, which can be used to systemically study the pathophysiology of human erythrocytes." He then explained existing research and detailed the limitations with current approaches:

> Our overall goal is to establish the integrative micro fluidic imaging technique to retrieve materials properties of single cell from membrane fluctuation, which can be used to systemically study the pathophysiology of human erythrocytes. Genetic and biochemical approaches have identified proteins that can be expressed by genetic defect (HbSS; sickle cell disease) or secreted by exogenous parasites (RESA, KAHRP; P. falciparum), and consequent changes in blood level have been studies in terms of vascular occlusion, flow rate change, and deformability. However, the connection in cellular level resulted from genetic level which can affect macroscopic change in blood level, or modified by local environment in blood which can alter biochemical mechanism intracellular level, has remained elusive. The limitations in our understanding of erythrocyte-related disease are, in part, because the causative or resultant changes in cellular material or mechanical properties have remained unexplored.

At the end of his opening paragraph, he provided a hypothesis, explained his previous research on the subject, and offered a clinical application: "A clear understanding of the direct effect of specific proteins to the cellular material properties will be fundamental to the developing prevention and treatment of those diseases," which he used to transition into his aims:

[Aim] 1. To determine the role of *P. falciparum* secreted proteins to material properties of human red blood cell. *P. falciparum* infected erythrocyte will be incubated in the micro fluidic channel and membrane fluctuation of the cells will be quantified to retrieve material properties: viscoelasticity of membrane cortex and viscosity of cytoplasm as well as hemoglobin concentration. The effect of each protein will be examined by knocking out the specific gene and confirmed by knocking in. The effect of RESA, MESA, KAHRP, and PfEMP3 proteins to changes in the material property will be studied. The specific binding sites of these proteins will be investigated by selectively destructing the possible cytoskeleton proteins. The effect of temperature (room, physiological, and febrile temperature) will be also systemically controlled and studied for each protein.

2. To investigate the role of sickle cell gene on material properties of erythrocyte.

3. To study the mutual effect of both sickle cell gene and *P. falciparum* secreted proteins in terms of malaria morbidity and mortality.

In writing these aims, Park had picked up on several concepts from the aims writing workshop. For example, he divided his proposed research project into three aims. He had also recontextualized his research to the point of view of a team of researchers rather than as an individual graduate student.

In responding to Park's draft, Ai had many technical questions about his measurable aims. She questioned how Park would obtain samples, what he hypothesized finding, and how he would conduct certain experiments. Ai was also concerned that the third aim was entirely dependent on the success of the first two aims. Although the third aim should be the riskiest, it should not be so dependent on the previous aims that if the research on the first two fails, the project cannot continue. Ai's main technical concern was that Park had "a lot going on" in his proposed research based on his aims. She wrote in her comments: "First you seem to be proposing a new imaging technique, which will provide in-depth information on material properties of diseased blood cells. . . . You also seem to hypothesize that altered mechanical properties (due to genetic or parasitic factors) lead to higher incidence of vascular occlusion (and that your device and approach can unique detect this?)."

In addition to issues of scope of his proposed research and technical details, Park seemed to struggle rhetorically with a sense of audience, as evidenced in the density of his prose. Park's writing not only had markers that English was not his first language; it was also laden with technical jargon. Ai, however, addressed only the technical jargon in her comments. She asked Park to clarify the "gist" of his research so that a broader scientific audience could read the proposal. She also asked him to add other rhetorical moves to contextualize the work in a broader framework. For example, she encouraged him to "spend a few more sentences describing how these diseases affect people, how they are currently understood/not understood/treated unsuccessful, how your research will advance current knowledge and/or treatment." In encouraging Park to add clinical significance to his work, Ai was encouraging him to move beyond the limited identity of a graduate student to a professional identity in which a researcher can identify the contribution his or her work makes to the field.

Park responded to Ai's feedback favorably. He commented that initially he was "too ambitious to organize [a] strong story" for his proposal, meaning that he had proposed more research than was feasible for this grant proposal. In Park's words, he "tuned it down a little bit" to make the proposal less ambitious. He also commented on the disjuncture in his aims: "These different specific aims are not connected. I realized that maybe I needed to organize, and I needed to focus what is really important, and what is really crucial in the topic of this proposal."

By the end of the semester, Parks' grant had gone through multiple revisions. With feedback from a communication instructor, he had also worked on some of the English as a Second Language markers in his writing. Although the final proposal was still highly technical, it lacked the overly dense jargon of his first draft (box 3.3).

Park's revised paragraph shows that by making his writing more accessible to readers, it is now clear that the overall goal of his research is to develop an imaging technique to study RBCs. The second and third sentences provide a rationale for that goal, and the final sentence of the paragraph signals to readers that this "niche" or void has not been sufficiently filled by other researchers. In four sentences, Park has told his readers what he proposes to study, the technical rationale for his work, and the place of his work in the field. He uses terms such as "overall goal" to signal key concepts to his readers and uses transitions such as "despite" to acknowledge the work that has been done in this area as well as what work is still lacking.

Subsequently, Park revised and reorganized his aims. In his revised aims, he proposed to develop a mathematical model of RBCs' properties, and then compare the results of normal of RBCs' membrane fluctuations with those found in malaria (*P. falciparum*) patients at different stages of infection. The most ambitious part of his proposal was still his third aim—to take RESA antigens that help hold together the cytoskeleton of RBCs and knock out various proteins to study their effects on the cells—but it was no longer dependent on the success of the first two aims. In writing the background section to his grant, Park's rhetorical task was twofold: to expand on his previous research and add a clinical application for his research. This was not an easy task for Park, and he struggled through several drafts to find a workable organization for his background and significance section. He initially organized this section such that it started with a technical explanation of the biophysics of RBCs, followed by an explanation of parasite-induced malaria, and then a description of the techniques to probe material properties of single cells, optical measurement techniques, and mathematical

Box 3.3

Park's revised aims for his grant, "Optical Measurement of Material Properties of *P. falciparum*-Infected Human Red Blood Cells"

Our overall goal is to establish a non-contact optical imaging technique to retrieve material properties of single red blood cells (RBCs) from their membrane fluctuations. These properties can be used to systemically study the pathophysiology of human malaria disease. During intra-erythrocytic development, the malaria-inducing *Plasmodium falciparum (P. falciparum)* parasite exports proteins that interact with the host cell membrane and spectrin cytoskeletal network. Parasite-exported proteins modify material properties of host RBCs, resulting in altered cell circulation. Despite the genetic and biochemical approaches identified, proteins exported by *P. falciparum* have remained elusive as well as the mechanism and effect of these proteins on the host cells. . . .

1. *To develop a mathematical model to retrieve material properties of red blood cells.* The dynamics of membrane fluctuation of healthy human red blood cell will be quantitatively measured by the optical interferometric technique. Correlations of membrane fluctuation at the specific distance will be calculated and will be combined with our mathematical model of red blood cells such that we can retrieve material properties of red blood cells: shear modulus of spectrin cytoskeletal network, bending and bulk modulus of lipid bilayer, and viscosity of cytoplasm. We will compare these results with the values probed by conventional techniques such as micropipettes and optical tweezers. Compared to the conventional techniques that probe material properties of cell, this method is expected to provide whole material properties of RBCs, monitoring a large number of cells systemically and non-invasively.
2. *To determine the role of* P. falciparum *in human RBCs in terms of infection progress.*
3. *To study the mechanism and effect of* P. falciparum *exported RESA and KAHRP protein to human RBCs.*

models for membrane fluctuation. With Ai's feedback "to guide the reader through the entire plan in logical, progressive manner," however, Park reorganized the background and significance section to more closely parallel his aims. He also foregrounded the clinical applications of his work.

In revising the background and significance section, Park said that he thought about "the big picture." In describing how he revised his work over the semester, Park reflected: "At first I was too ambitious . . . in terms of telling the story or telling the big picture. And during our first two [revision] processes, I realized that the way I presented was not understandable. So basically I just wrote again from the script. That really helped to reorganize the structures of this whole plan."

Reviewers responded favorably to Park's proposal, and he received a high score from his study section reviewers. The primary reviewer commented that Park's proposal would "provide a cost effective means to track malaria infection on a cellular level, by understanding its precise effect on the material properties of infected cells.... Overall, this proposal is well organized and clearly planned." Park was particularly pleased that the reviewers understood the novelty of his ideas and "they appreciated why this is important and why it is very crucial in approaching this kind of project."

His reviewers, however, had criticisms of his work. In addition to wanting even more information on the significance of his research, the primary reviewer commented on the grammatical errors in Park's writing: "There are many missing words and grammatical errors in the background section—should proofread. (e.g. "every year, it causes about over five hundred million people" or "RBCs fail to the splenic sinusoids ...", etc)." Grammar and style are not in the NIH scoring criteria, but it is widely known among professional researchers that reviewers often score proposals based on such criteria. Park's struggles with grammar marked the most difficult challenge for him in the Frontiers class. In our final interview, he reflected on the reviewer's comments regarding his grammar: "I had a really difficult time writing, should I write *a* or should I write *the*. I was frustrated when I got the other people's review of my grant proposal ... when they picked up the mistakes. Even though I [threw the reviews] away, I [felt] good ... I think I have to use this kind of critique."

In reflecting on the reviewers, Park tried to take the criticism constructively, but his frustration was apparent. Grammatical correctness is not a trivial issue for students like Park. Although scientists are often very tolerant of World Englishes in spoken language, they are less accepting of variation in writing. In fact, although the contemporary practice of science is global, its written forms do not reflect the diversity of its practitioners. It is assumed that students like Park will be able to perform standard academic English flawlessly with little support while they are in graduate school or later in their careers. Yet to focus only on the grammatical mistakes in Park's writing misses the deeper developmental changes he demonstrated. With Ai's feedback, Park was able to conceptually rethink the organization of his ideas (without the help of his mentor) such that they fit the logical expectations of NIH reviewers. Many of the other Frontiers students struggled with that concept, even with the help of their mentors. Park also was able to relinquish the technical jargon of his earlier drafts, and by the end of the semester, he was crafting prose that was appealing to a wider range of scientific audiences.

Yet Park's reviewers did not see these changes since they read only his final document. If reviewers go beyond the rubric, they can likely sway other members of the study section to score accordingly. Park, like Joe, commented on this issue: "Most of the opinions of the [study section] reviewers are driven by the first main presenter. I had experienced this during our mock session because I was reading as a second reviewer, and I thought that the idea was very novel, but after hearing that the first reviewer was saying . . . I come to follow his ideas."

The benefits of Park's hard work all semester were short-lived. At the end of the semester, he had not convinced his mentor to take an active interest in this particular project. Because his work was not the lab's main focus, it was unlikely that the lab would spend many resources to pursue this research further. By the end of the semester Park had changed his research focus to suit the interests of his lab group. Although he was still working in optics research, his newer work focused on early cancer detection. Nevertheless, he remained hopeful that his mentor might be interested in the RBC work given the progress he had made on the grant. In our final interview, Park concluded, "Maybe I'll talk to my advisor."

Summary of Frontiers in (Bio)medical Engineering and Physics

The case studies of Joe, Alyssa, and Park complicate notions of school and work, as well as notions about collaboration. In learning to define and motivate their own research, graduate students in biomedical engineering find themselves at the edge of their student identities: they are almost professionals and seek a professional identity, yet because of the material resources they require to engage in scientific practice, they remain under the wing (or in the shadow) of their mentors. The studies of these three students also illustrate that the genre of the grant proposal in itself was formative to students, as was learning about the NIH review process. Perhaps most important, this chapter shows how modeling professional peer review can contribute to student understanding of scientific writing as a persuasive endeavor with the successes and pitfalls of the peer review process. In sum, the main themes that can be taken away from this study of graduate students' grant writing are as follow:

Learning to define a project and to organize the literature of the field was an iterative process. While students can learn the persuasive nature of scientific communication through research articles or other genres of scientific writing, grant writing was particularly useful in teaching students how to motivate their research. First, the grant proposal genre pushed students to organize information in certain ways. The aims section requires that writers distill their proposed research to a single page and be able to chunk the proposed research into three separate aims, and the background section requires writers to contextualize and organize the vast literature of a field. The tightly constrained format of the grant genre was itself meaningful only if students understood how readers approached grants. Through Bhatia's insider perspective as an NIH reviewer and Ai's feedback, students repeatedly were encouraged to move beyond technical descriptions of their research. Through multiple opportunities to reformulate their sales pitch, students came to a better understanding of the subtle rhetorical devices scientists can use to sell their research. As Ai explained, these iterations pushed students to "start really looking at what the field is, which is critical to understanding what you're improving."

While the grant writing activities took place within the space of a classroom, the influence of mentors was a mediating force on student learning. The goals in this class were about how to organize a project, define what is important or significant in the field, and convey that information to readers. A second dimension to the class then complicated our understanding of student learning. The message that the PIs sent to students about their audience's expectations, the ethos of a professional scientist, or the direction that their research should take all had marked impacts on the Frontiers students' grant writing. In Alyssa's case, the mentor's input led to tensions. In Joe's, the mentor's input helped push him forward in his research. In Park's case, his mentor did not openly work against the instruction in the class but gave Park few opportunities to apply his new knowledge. Bhatia did not see the impact of the mentors as a problem. In her own role as a mentor, she saw it as a "larger truth of graduate school": "If your project is at the center of gravity of the mentor's interest, you get more attention, more positive feedback, more interactivity, or the mentor uses your slides. There's much more of a dialogue around your work than if you work on a peripheral project."

Engaging in authentic forms of peer review helped students better learn about the persuasive strategies used in grant writing. Overall, in terms of asking how authentic writing activities like grant writing contribute to student learning, we can say that authentic learning needs authentic feedback and authentic peer review.

As the case studies throughout this book show, learning to communicate like a professional requires faculty, mentors, and teaching assistants to respond to student writing in ways that enact professional expectations. Ai, like the faculty profiled in the next chapter, responded to student writing in ways that modeled professional expectations. In the case of Joe, Ai's feedback was also complemented by feedback from Eli, Joe's mentor. Such feedback brought out a reflectiveness in Joe, which ultimately produced further growth (and a grant submission).

But feedback from teaching assistants and mentors was just one element of the Frontiers curriculum. The most important aspect of the class in terms of student learning was the mock NIH study section, which modeled authentic peer review (even though the criteria were truncated). The authentic peer review of the study sections not only put students in the role of professional reviewer, it also put students in a different receiving role. Students overall took up the responsibilities of reviewer seriously, and they responded to their reviews in much the same way that professional scientists respond when receiving NIH scores. Students took away from this experience a deeper understanding of the constructed nature of scientific peer review. As Bhatia explained, the peer review process teaches students to become conscious of their own biases as they enact the role of reviewer:

> In the academy, we have opinions about different bodies of work, because we have our own internal biases. And those come together in our minds to give a collective evaluation of the piece. That's exactly what you have to go through in study session. Students come to recognize that in their own evaluations. When they're not excited about a project, that's because they didn't think it was significant, or the student didn't motivate it sufficiently. So when you were the reviewer, and you respond viscerally to something, it teaches you about your [own biases].

In the end, students were spared perhaps the most constructed aspect of the grant proposal process: the funding cutoff. As Ai explained, under current NIH funding guidelines, the cutoff for the Frontiers pool would have been perhaps two students. Bhatia and Ai decided not to give students their final rankings because, as Ai explained, "I do not want to tell you that you were the last grant."

4 Learning to Argue with Data

The idea that a scientist gets up and dispassionately tells you discoveries that he's made were completely unmotivated [and he went into the] situation with no bias. That's utter garbage. Everybody's out to prove something, and you may not admit in your final presentation what you were working on initially, but you have to pitch it within some context or people just will not be interested.

Dennis Freeman, professor, MIT Electrical Engineering and Computer Science

In this chapter, we look at the ways in which students are taught to argue like scientists using data—how to look at the raw data to find an interesting finding, how to analyze and present that data in ways that extrapolate the interesting finding from the raw data, and how to convince their readers that the data are meaningful and accurate. The case studies in this chapter explore the following questions:

- How do students learn the persuasive devices that professional scientists use when communicating data to other scientists?
- What challenges do students encounter as they learn to use visual evidence in scientific communication?
- What role does faculty feedback play in the development of this professional skill?

The role of evidence and argument in writing and speaking is well articulated in the humanities and in professional fields like medicine. (For reviews, see Monroe 2002, Segall and Smart 2005, Bazerman and Paradis 1991.) But the public face of science as a dispassionate, objective exercise in recounting facts often overshadows the complex, often difficult choices scientists make in designing experiments and reporting data. For students, such misconceptions come to the fore as they engage in the activities that professional scientists encounter the same kinds of dilemmas practicing professionals face: Is there a finding in the data worth publishing? What data should be presented? How much can the data be shaped without seemingly being unethically manipulated? What formatting techniques can be used to best make the visual findings readable and clear? In answering such questions, students discover that the representation of scientific data goes beyond simply making aesthetically pleasing figures. In learning how to select, shape, and present data, students learn the ways that professional scientists make arguments with quantitative evidence, and they also are introduced to the constructed nature of the scientific research process.

One important dimension in scientists' selection and presentation of data is their understanding of how those data will be received by the scientific community in which they work. More than thirty years of research from sociologists such as Bruno Latour (1979, 1987) have shown that whether weighing evidence, interpreting data, or assessing risks, scientists make daily decisions based not on purely factual data but on how those data will be received by other scientists (Lynch and Woolgar 1990; see also Pauwels 2006 for an overview). This is not to say that scientists make up their findings or do not deal with factual evidence, but what they choose to observe, their selection of methods, the data to analyze, and how to represent and report those data are all wrapped in human choices that anticipate how other researchers will receive those data (Lynch 2006b). Thus, the community of practice in which scientists work is a powerful motivator in the selection (and sometimes manipulation) of data (Knorr-Cetina 1999; Resnick, Shamoo, and Krimsky 2006). What ends up being presented as visual evidence in the final published research article is a reorganization of raw data in support of particular claims (Lynch 2006a).

Amann and Knorr Cetina (1988) distinguish between the *raw data* gathered in the lab and the *evidence* presented in scientific publications. The data that are collected in the experimental procedure—notes, random jottings, messy plots, massive computer outputs, and other visual traces—are the artifacts of questions pursued, questions left unanswered, and experimental mistakes (Lynch and Woolgar 1990, Baigrie 1996). Data become evidence only "after they have undergone elaborate processes of selection and transformation" (Amann and Knorr Cetina 1988, p. 88). McGinn and Roth (1999) call this *sanitization*: the process of moving from physical specimens to raw data to graphs and plots "that can be shared with and interpreted by other scientists" (p. 20). In this process, only some data end up as evidence in scientific papers or one of the visual genres scientists use to communicate with other scientists, such as the plot with error bars.

In *Shaping Written Knowledge*, Charles Bazerman provides an example of this process in Arthur Holly Compton's empirical verification of quantum theory in 1923. Bazerman describes how Compton selected data from some experiments but not others to report and that the data tables in Compton's notebooks "are filled with corrections"

(1988, p. 208; for other examples, see Knorr Cetina 1981, Prelli 1989). To retain the "integrity of the data" and ensure readers that he is "constrained by the data and not fiddling with it," Compton reports the error in his calculations while leaving the equipment and natural phenomena out of critical reach (Bazerman 1988, p. 209). There must be the appearance that the visual evidence stands on its own without the researcher, yet the researcher's influence is always present in what is delivered to readers.

In scientific research articles, visual representations of data are the work horses of argument. Indeed, visuals comprise, on average, 26 percent of the surface area of the twentieth-century research article, with the Cartesian graph being the predominant visual found in research articles today (Gross, Harmon, and Reidy 2002). In the research article, the visual ultimately becomes a tool "for reasoning about quantitative information" (Tufte 2001). Error bars or selective labeling may be added to "clarify, complete, extend, and identify conformations latent in the incomplete state of the original specimen" (Lynch 1990, as quoted in Gross et al., p. 206).

Yet visuals ultimately do not speak for themselves (Mathison 2000, Bean 2008). In their transformation of data, scientists complement the visual evidence with textual cues to help direct a reader to the desired interpretation of the evidence. The text works in conjunction with the visual to become a "work site" where the author puts in words the argument to be made in the accompanying figure (Roth, Pozzer-Ardenghi, and Han 2005). In other words, the text accompanying the visuals tells the reader what to see in the image. The figure reference acts as a portal from the textual to the visual argument and moves the reader's eye from the text to the visual and back again.

Students struggle with the representation of data on multiple levels. At the most basic level, they must learn how to use laboratory equipment for generating the data. Once they have gained some fluency with the use of equipment and how to output data, they often find themselves at a loss of what to do with the data tables and figures they have produced. At this point, they are likely to simply dump raw data tables into laboratory reports with the claim that the data support a given theory.

Moving students beyond this moment to think about the visual evidence in scientific arguments is challenging for several reasons. First, students are usually not privy to the process of selection that professional scientists use in choosing data for publication. As Natalie Kuldell pointed out to her students in chapter 2 of this book, researchers decide what evidence to present in a scientific research article based on their confidence in the data. Scientists gather multiple kinds of data to check the veracity of their findings. Only when they have confidence that the findings are accurate do they present the data. Findings that cannot be repeated are unlikely to get reported. Second, part of students' misconceptions about the function of visual representations comes from textbooks. Scientific textbooks are filled with accepted visual models of scientific phenomena, most of which are plotted using smooth curves with no error bars and few data points. The text supporting the plots describes visual representations as proof of a phenomenon. The collection and analysis of scientific data are presented as methodological choices that are the result of purely scientific best practices, not what methods will yield the kinds of data that most interest a particular scientist and will persuade readers. Third, textbooks on scientific writing focus on the "correct" presentation of tables, bar charts, and line graphs, truncating the process that yielded that visual evidence. (Even well-known handbooks have this problem. See Perelman, Paradis, and Barrett 1997; Gurak and Lannon 2001. For useful discussions, see Porush 1995, Penrose and Katz 1997.)

At MIT, like other universities, students take up these miscues in their writing up of research findings. They often fill their reports with hastily drawn plots, offer sparse textual support of their figures, or include aesthetically pleasing plots that ignore most of the actual data that they gathered. As the case studies in this chapter show, laboratory exercises allow students to design and conduct their own experiments and give them the opportunity to learn how to make sense of data: how to select, interpret, explore, present, and receive feedback on data.

From a teaching point of view, the challenge is to teach students about the kinds of rhetorical moves that scientists make with data without resorting to fill-in-the-blank templates or misleading students to believe that all decisions in science are relative—both of which diminish students' respect for the scientific process. Helping students understand and participate in the logic that girds scientific decision making requires addressing students' misconceptions about the ways scientists use visual arguments while avoiding accusations regarding the unethical manipulation of data.

In this chapter, we address how students learn to work with visual evidence. In teaching undergraduates how to make visual arguments with data, our goal is to help them integrate the sense-making process of analyzing data with audience expectations for visual evidence.

Focusing on Visual Evidence in Quantitative Physiology

Quantitative Physiology: Cells and Tissues (from now on called Quantitative Physiology) is a large-lecture undergraduate course in biomedical engineering. The course is housed in the Department of Electrical Engineering and Computer Science (EECS), the largest department at MIT, and is taught by EECS faculty. Although the department has many traditional electrical engineering courses as well as computer science offerings, it also has integrated biological and nanotechnology research into the curriculum. EECS has cross-listed many of these courses to appeal to students in biological engineering as well as other engineering programs that have also become more interdisciplinary, such as material science and mechanical engineering.

Many of the thirty-one students in Quantitative Physiology during fall 2007 were junior and senior biological engineering majors or EECS majors who have an interest in biomedical research. In the class, students learn the principles of mass transport and electrical signal generation for biological membranes, cells, and tissues. In addition to lectures and recitations, students complete two lab experiments: an experimental microfluidics project in a wet lab and a computer-simulated study of the Hodgkin-Huxley (HH) model, a scientific model based on nonlinear differential equations of how action potentials in neurons are initiated and propagated.

Quantitative Physiology carries a communication-intensive designation and is considered an integrated model of communication-intensive (CI) instruction because the communication and technical instruction are intertwined throughout the semester. The course is supported by MIT Writing Across the Curriculum (WAC) program instructors who help design assessments of student writing and speaking, team-teach communications lectures, and provide feedback on student writing. The course faculty, who have worked with the MIT WAC program for more than ten years, make an explicit connection between communication and content. As stated on the course syllabus, "This subject is communications intensive. We feel that communications skills are essential for professional engineers and scientists. We also feel that the process of creating written manuscripts and oral presentations can help clarify thinking and can be an effective way to *learn technical material*" (6.021J Course syllabus 2007, emphasis in original).

Jongyoon Han, the EECS professor responsible for Quantitative Physiology in fall 2007, explained the faculty's reasoning behind linking technical content and communication: "Studying the most effective way of communicating your technical idea to the audience or readers reinforces your understanding of the very problem. In the course of doing communication activities or practicing those skills, I have found that student understanding of their problem actually deepened further in the practice of trying to communicate to others; they actually ask the questions that others would ask."

With the exception of the writing clinic and a communications lecture, all activities in class were completed outside normal recitation and lecture time. Students began each lab project by submitting a research proposal. After the proposal had been accepted, they completed their data gathering in the lab and drafted their report. The report underwent a parallel review process by three readers. Students met with reviewers at a writing clinic to discuss each other's comments. Following the clinic, students revised and resubmitted their report. The entire process, from proposal to final report, took four weeks.

The communications curriculum in Quantitative Physiology was developed out of a perceived mismatch between students' research skills and their ability to communicate those findings. In discussions, faculty noted that the source of student writing problems was in their approach to the data. Students attempted to present all of their data rather than a meaningful subset; they offered limited or no explanations of their findings; they did not know how to use data throughout the article to make a convincing case for their work; they provided no coherence across the sections of research articles; and they seemed more interested in forcing the data to an acceptable theory rather than focusing on interesting findings. With this in mind, Dennis Freeman, the first faculty member to suggest adding writing instruction to the course more than a decade ago, suggested the use of storyboarding, a technique used in filmmaking to help filmmakers visualize the entire narrative in a film before shooting begins.

In incorporating the storyboarding approach, Quantitative Physiology faculty identified three underlying concepts to storyboarding that they wanted students to learn:

1. Data drive scientific research. Organize and locate trends in data before beginning to write the supporting text.
2. Each figure in a report tells its own story. Design figures that make the point that you want to make.
3. In sum, the figures in a report tell a narrative of the research. Consider if the data make a logical sequence from one figure to the next figure.

The storyboarding concepts were not just explained in class. In Quantitative Physiology, story boarding and data analysis were criteria for peer review and evaluation of drafts. They were also worth 30 percent of the report and presentation grade for each project. Following is the rubric for data analysis for the microfluidics project grade sheet:

Storyboarding (Selection of data)/Figures/Captions (10%)

A: Figures/Captions in the results section are well-prepared, clear, and key trends can be easily captured. No extraneous materials or figures were added.

B: Figures/Captions can be improved to increase the clarity. No extraneous materials were added, and all the figures shown are helping to make the conclusion.

C: Figures/Captions leave something to be desired, and extraneous materials were shown without serving any good purpose.

D: Figures/Captions are not organized or processed and presented as is.

Data Analysis/Results and Discussion (20%)

A: Analysis of experimental result is free from technical error, and the results convincingly support the arguments/conclusions made in the report.

B: Analysis of experimental result is free from technical error, and the results are consistent with the arguments/conclusions made in the report.

C: Analysis of experimental results has minor technical errors, and/or the conclusions drawn are not supported by the data presented in the report.

D: Major technical errors or too little technical content or too poorly written to assess technical content.

The emphasis in Quantitative Physiology on data is central to creating an authentic research exercise. In the first class project, the microfluidics experiment, students were to design and conduct an original research study using microfluidics devices. Students were given mouse cells, which were placed in microfluidics chambers. The chambers allowed the cells to be "trapped" while various solutions were added to the chambers, and the cells shrank or expanded in response. A computer program, CamScope, then took pictures of the cells as they changed.

The primary technical goal of the project was to give students insight into osmosis and diffusion. The stated learning goals for the project were "designing an experiment" and "acquiring, processing, and interpreting experimental data, and communicating the results to others" (Microfluidics Project Overview 2007). Students were told explicitly to "make self-consistent measurements and provide a rational explanation for those measurements, not to validate or refute a particular theoretical model" (Microfluidics Project Overview 2007).

In regard to the final product—a written report—the Quantitative Physiology faculty suggested that students start with a storyboard of their data to help them develop a logical argument of their results:

> We strongly recommend that you start by developing a storyboard to structure the logic of the report. Revise your storyboard until the flow of your argument makes sense (ask yourself, "If this were someone else's report, would I believe the conclusion?").... Generally, results can be communicated more efficiently and accurately with figures and graphs than with words alone. However, a collection of graphs without a written description of their relevance is not acceptable. The text of the results section should carry the reader through a presentation of facts that is intended to lead to a conclusion. (Microfluidics Project Overview 2007)

In the second class project, the Hodgkin-Huxley experiment, students were to design and conduct an original research study using a computer model of squid nerve axons and then deliver a final presentation of their results. The computer model provided students an opportunity to learn about the complex physiological behavior of cells in a controlled context. In the computer simulation, students could alter the biological values associated with action potentials (i.e., the firing of nerves) and then determine how sodium and potassium altered the nerve's response. As in the microfluidics project, students were not out to prove a hypothesis in the Hodgkin-Huxley project: "The Hodgkin-Huxley model is sufficiently complex that investigation of any of the hypotheses will most likely lead to unexpected results. You should pursue these unexpected results and try to understand their bases.... Your aim should be not simply to reject or accept the hypothesis but to delve into the topic in sufficient depth so as to deepen your understanding of the model" (Hodgkin-Huxley Project Guidelines 2007).

As the instructions for both projects make clear, students in Quantitative Physiology were to explore unexpected results in their research. This suggests a different kind of approach to research than students normally are offered in that they are typically expected to provide the "right" answers to scientific problems. By assigning plausible rather than "right" answers to their findings, they were challenged to draw on their persuasive abilities to convince readers that their interpretations were both accurate and interesting.

Learning to Argue with Data

The two case studies in this chapter are based on two students who enrolled in Quantitative Physiology in fall 2007. For these studies, we chose to focus on students who were further along in their development and whose stories tell us about the evolution of student learning as they attempt to approximate the activities of professional scientists. Their stories, along with those of three other students who are not presented in this chapter, show how students move along a developmental continuum in learning to use data in making scientific arguments. In the case studies not presented in this chapter, we found two students wrestling with an understanding of what data to gather and how to make sense of those data in the most basic of ways. Their stories are important, but they do not tell us about the furthest trajectories of student learning that build on the case studies we presented in the first three chapters.

In addition to the case study data, we gathered survey data from seventeen other Quantitative Physiology students. Only about 40 percent of the students had any instruction in communicating scientific findings, and none of them had read a professional scientific article in the previous year. In pretest surveys, students said that the results section of a scientific article should include all of the data gathered during experiments. By the posttest surveys, this misconception had been corrected. Also, while students had misconceptions regarding the scope of the data to be presented in a scientific article, they understood that figures needed captions and that the discussion section of an article needed evidence to support interpretations of the data offered. Finally, although students had mixed understandings of the scope of data offered in a scientific research article, they did have a sense of the importance of persuasion in scientific communication. In their pretest surveys, students wrote the following when asked about the importance of argument in scientific writing:

"Data can be perfectly correct, but analysis is what is subjective. Argument gives readers something to consider and can be argued if results are interpreted differently."

"Only in discussion. This is where you argue for your interpretation of the data. Without this, your data do not mean anything."

"The writer of the scientific paper must argue convincingly to persuade the reader that his experimental results do support the claim that he makes in the hypothesis/discussion sections. For example, if he claims that A causes B, he must convincingly argue that B is not, in fact, caused by some other C, and that he didn't make some mistake in his experiment."

These responses reveal several important concepts. First, students understood that scientific communication is not merely a recounting of factual details. Second, students made a distinction between the data gathered and the interpretation of those data. They did not consider methodological choices used in gathering data as a series of rhetorical choices. Persuasion is left to the interpretation of data, not its collection.

At the end of the semester, we asked students to reflect again on argumentation in scientific writing, specifically about the visual display of quantitative evidence. Once again, they thought about scientific writing as rhetorical, but they came to a deeper understanding about the relationship between audience and reception and the presence of persuasion throughout the scientific process:

"I realized even though facts and figures are an important part of scientific writing, presentation of these data play an important role toward acceptance [of your work]."

"You may have to argue for the veracity of your methods and your conclusions, far more than I thought at the beginning of the semester."

Together, the case study data and survey results point to three major findings regarding how students learn the persuasive devices professional scientists use to communicate data:

- Students follow a developmental trajectory that begins with talking about what they are doing methodologically and how they are doing it, to why certain data are interesting and what the limitations or possibilities of that data are.
- Different communicative genres lead to different learning experiences in understanding how to use scientific data.
- Faculty feedback that models authentic, professional feedback helps guide students in developing their understanding of using data in making scientific arguments.

In the following case studies, we take these findings and examine them in more detail through the stories of two students: Devi and Julia. For Devi and Julia, the overall experience in Quantitative Physiology was formative. Devi had already gained basic proficiency with scientific writing but had not conducted original research and struggled with finding the right story in her data. Her initial attempts at making sense of the data led her to offer an unconvincing argument to her technical reviewers. The discussions that follow show how visual evidence became the focus of discussion in scientific arguments while the raw data and approach remained untouched. For Devi, Quantitative Physiology brought new insights into how to make sense of data and ascribe a compelling narrative to those data.

Julia had an advanced understanding of data gathering, selection, and presentation. For Julia, the focus was not so much on what to present in her data or what story to ascribe to those findings but how to work with problematic data in ways that still yielded useful results. She also attended to the detailed aspects of data presentation, such as formatting and coloring. Thus, for Julia, we see an awareness that scientific data can often be murky and an evolving sense of how professional scientists can have confidence in data that are less than ideal. Quantitative Physiology also confirmed her desire to situate her work within a larger professional context.

Case Study 1: Devi—From Visual Miscues to "Interesting" Data

Devi, an EECS junior, entered Quantitative physiology having never taken a class that taught scientific writing and having never given a technical talk. Her survey answers showed that she was not familiar with other expectations for scientific articles—for example, that the results section does not need to report all of the data gathered in the lab experiments. Despite her lack of experience, she did understand that figures were not self-explanatory in scientific articles and that scientific communication included persuasion. When asked about the purpose of a scientific article, her response also showed that she had a good bit of rhetorical awareness about scientific audiences. As she wrote, the purpose of a scientific article is "to show the capability of the student to conduct an experiment based on a hypothesis and present the results and conclusions in a written format that others can understand." In Devi's view, scientific papers are a demonstration of the researcher's ability as well as a reporting of findings.

In terms of the two class projects produced in Quantitative Physiology, Devi ultimately produced a well-received research article and a well-received oral presentation. She received a high grade for the course, as faculty noted her diligence and attention to detail. But the process of getting to final high-quality work was the result of an iterative process of working through her data. In this case study, we see how Devi struggles with finding a plausible story for her data in the first project and the importance of her visual presentations in that process. Later, we see that as her confidence as a researcher grows, she can move beyond finding accurate interpretations of data to exploring interesting findings and explaining those findings in a way that seems plausible to other scientists.

For the microfluidics experiment, Devi and Pasha, her lab partner, decided to measure the rate of cell swelling when subjected to varying concentrations of *Cytochalasin B*, a fungus toxin. They read that at high concentrations, *Cytochalasin B* breaks down the cell's internal structure, thus altering the cell's normal osmotic response. Devi and Pasha wondered if this reaction might make the cells react more slowly than under normal conditions, so they ran several trials in which they varied the concentrations of *Cytochalasin B* and observed the cells' responses. Turning to the graphical outputs, they could then begin to make sense of the microscopic changes they could barely observe with the naked eye. A graphical representation showed "some changes," in Devi's words, due to the *Cytochalasin B*. Devi and Pasha chose four graphs, grouped together to show the cell's change in cell diameter over time in response to varying concentrations of *Cytochalasin B* (figure 4.1). Each graph was a scatter plot and included a linear approximation (the straight line) in the region of cell swelling as well as a regression line (curving line) that had been fit to the data points.

In commenting on their results figure, Devi and Pasha described each figure but did not tell readers what each result showed:

> [Figure 4.1] illustrates the effect of *Cytochalasin B* on the osmotic response of the cell under 0.25x PBS solution for the duration of 8 minutes. The four graphs present in [figure 4.1] represent the change in diameter (in pixels) of mouse white blood cells exposed to 0 μg/ml, 5 μg/ml, 8 μg/ml, and 10 μg/ml of *Cytochalasin B* respectively. Each graph is a third degree polynomial regression based on the raw data obtained during the trials. Furthermore, a straight line was fit to the region of cell swelling to characterize the steepness of the transition. In order to obtain a better understanding of the spread of the raw data, a scatter plot will also be presented on the same graph.

In other words, Devi and Pasha told readers that the figures represented changes in diameter of mouse white blood cells instead of telling readers, for example, that the

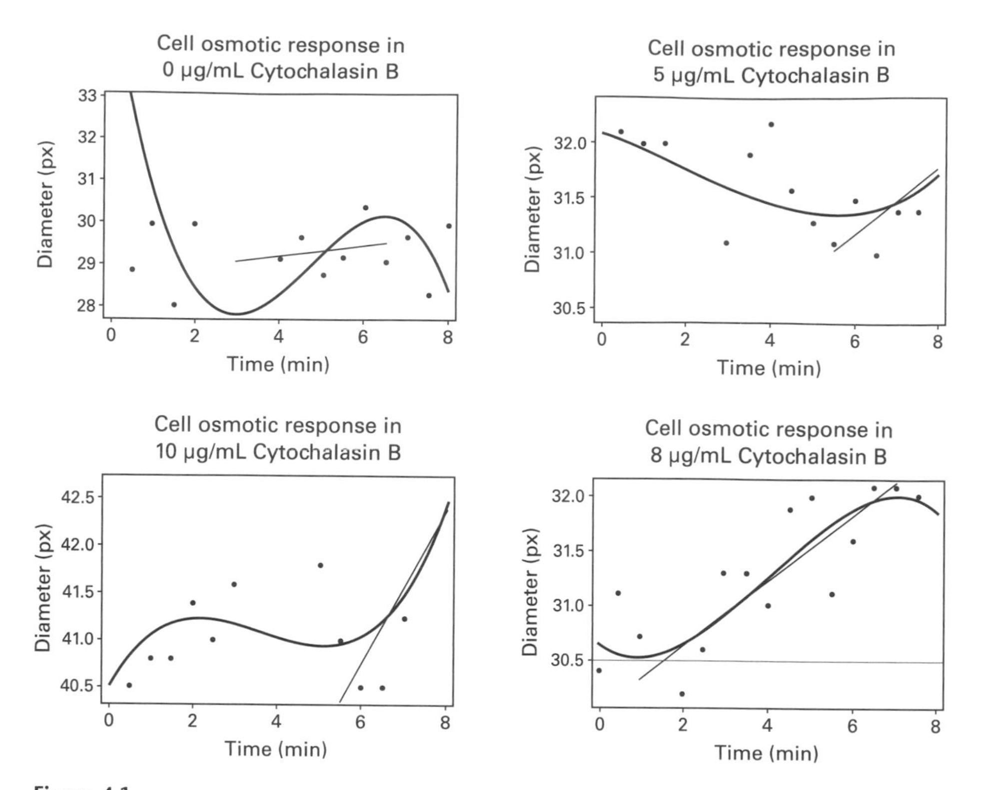

Figure 4.1
Results figure from Devi and Pasha's draft microfluidics report

cells typically shrank first before expanding. They also did not explain what readers should see regarding their regression analysis, although they did write that the lines are in "increasing order of slope." Instead, they placed their description of these findings in their discussion section.

In their discussion analysis, Devi and Pasha noted that the cells shrank initially, but they diminished the significance of this finding and directed readers to the point where the cells swelled. Cell swelling confirmed their initial thinking that the cells would expand in an osmotic response.

Reviewers—faculty, teaching assistants, peers, and communication instructor—were not convinced by Devi and Pasha's graphical representations or the supporting explanations for their findings. Sangeeta Bhatia, the faculty member who reviewed Devi and Pasha's draft, wrote:

> This is a good analysis but there are two main flaws. First, the data analysis and curve fitting do not seem valid. I could hallucinate any number of curves to fit your very first data set.... In addi-

tion, you state throughout that [*Cytochalasin B*] contributed to the disintegration of the cytoskeleton; however, you have no such evidence—at best you can say that your data are consistent with such a theory but you should consider other mechanisms of influence.

In this critique, Bhatia objected to both the way that the data were presented and the ambitious explanation provided for the change observed. In their presentation of the data, Devi and Pasha had replicated the smooth trend lines offered in the textbook plots. Bhatia, however, read the draft as a professional scientist who was unconvinced by smooth curves lines based on scattered data points. Her choice of the verb *hallucinate* to describe the curve fitting in the first figure strongly indicated her lack of confidence in the trends that Devi and Pasha reported.

Bhatia's second objection was that Devi and Pasha ascribed a biological interpretation to the changes shown in the plots without considering other factors that could have resulted in those changes. In other words, they overstated the causal force of the *Cytochalasin B*. In her comments, Bhatia reminded Devi and Pasha that such data do not "prove" a theory; scientists can say only that such data are consistent with theory. Data have limits because they are only partial representations of natural phenomena. In systems biology, many factors can lead to the raw data outputs shown in a plot. A sage researcher hedges sufficiently to account for other possibilities in the outcomes.

A third, and more serious, problem with Devi and Pasha's data became evident during the writing clinic, the recitation hour dedicated to talking to reviewers (faculty, TAs, peer, and communication instructor) about their comments. As Devi and Pasha talked to Bhatia and the teaching assistant, Hays, during the writing clinic, it became clear that something was wrong with the experimental data reported in the article. Rather than looking at the raw data to find the source of the error, however, most of the conversation between Devi and her reviewers focused on the figure with the smooth curve fitting. The problem unearthed through this conversation about the plots was that Devi and Pasha had not fully considered the process of osmosis. In this case, they had not considered that the cell might swell first, and so they returned to the lab for more experiments. They had, in fact, learned an important technical idea by presenting bad data. Devi and Pasha's revised results section retained the four-figure comparative presentation of the data as in the draft, but instead of plotting the cell diameter change over time, they plotted the change in the cell volume over time (figure 4.2). Like the previous figure, this one has four small plots, each representing a different concentration of *Cytochalasin B.*

The interpretation that Devi provided in the discussion section reflected her growing understanding of physiological changes in cells—cells' responses of osmosis and diffusion occur at different rates:

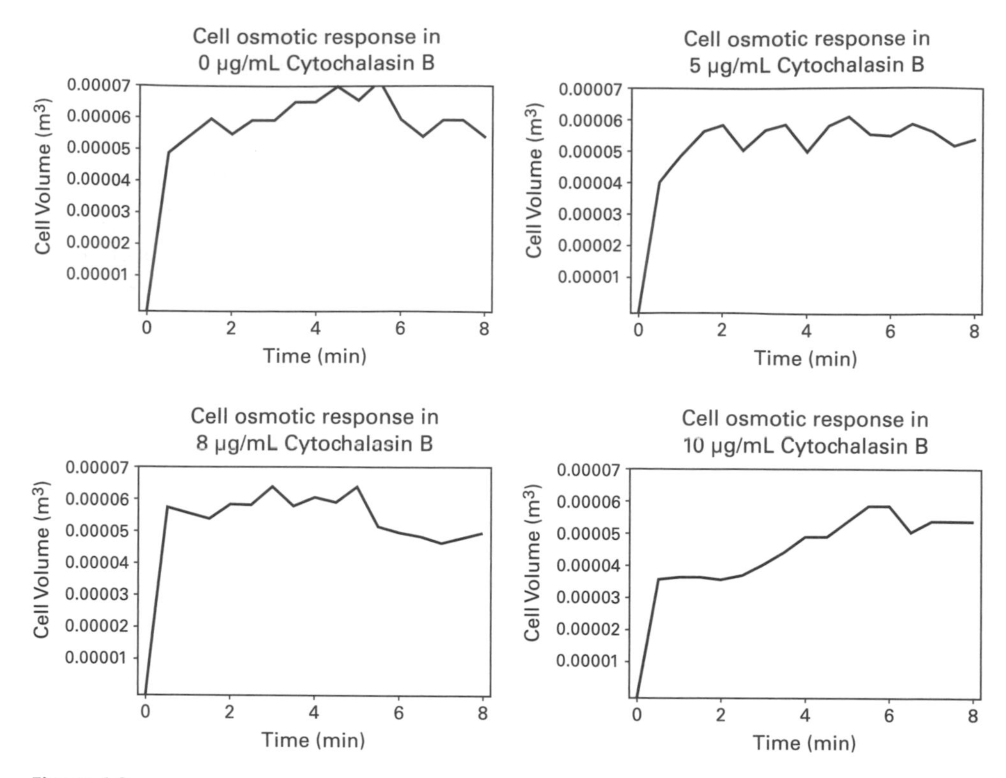

Figure 4.2
Results figure from Devi and Pasha's revised, final microfluidics report

Figure [4.2] shows the graphs that contain the volume vs. time representation for various concentrations of *Cytochalasin B*. As shown by lines connecting the data points, the cell volume behaves approximately exponentially with the change in time. This is consistent with established cell behavior in the face of osmotic shock as presented by Weiss in section 4.7.2 in the text. Since the osmotic response of the cell is much faster than the response of the cell due to diffusion we see that the cell starts to swell initially because of the concentration difference from the inside to the outside of the cell. As diffusion takes place the increase of the cell volume starts to decrease until it hits the peak value. When diffusion starts to dominate over osmotic response the volume starts to decrease and reach its equilibrium state.

Bhatia and Hays liked many of the changes in Devi and Pasha's final report, but they still had concerns about the figures. Hays wrote:

Good job trying to do more quantitative analysis. I think the analysis you were attempting to do was to compare the rate of change of cells under different *cyto B* concentrations and I believe therefore that the slope is what you were trying to capture in comparing the "peak volume change/time to peak." ... starting the line at cell volume 0 at time 0 is misleading. I do not see that [your figures show] the swelling and then shrinking behavior "explicitly." . . . Most of the data shows an insignificant effect of the drug.

In his final comments, Hays rewarded Devi and Pasha for making their analysis "more quantitative." His concerns about the data, however, were quite different from the concerns on the report draft. While his draft comments focused on major interpretations of the data and the overly ambitious curve fitting, his final comments focused on the presentation of the data themselves in the report and some claims that needed further muting. Hays pointed out that the axis labeling was misleading and that the data showed less of an effect than what Devi and Pasha claimed in their supporting narrative. In the end, although he still had concerns about the final report, it was clear in Hays's final comments that Devi and Pasha demonstrated a more convincing story of their data than in their report draft.

In reflecting on her experience with the microfluidics experiment, Devi was animated in describing how her increased understanding of physiological mechanisms came about through conversations with her reviewers. Through those conversations, Devi noted something important about scientific review—that reviewers do not "know" if data are correct: "I think all of them . . . thought that we understood our own experiment so they were trying to interpret [the data] that were [represented]. All of them had this problem, which was why are you looking at this increase? What is that supposed to mean? But they didn't come out and say look, we think you did the experiment wrong."

Devi had learned that not only do data not speak for themselves, but that they can tell lies. Scientific reviewers look skeptically at data, but they also assume a certain level of technical proficiency on the part of the researcher. Reviewers can base their judgments on only what is presented in the paper. This experience challenged how Devi thought about the relationship of scientific audiences to data and made her more attentive to her responsibility as a researcher to work through interpretations of data before publishing them. This experience would be useful for the the Hodgkin-Huxley project.

For Devi, the primary challenge of the Hodgin-Huxley computer simulation was in understanding the anomalies in the data, which were to be explored in more depth. Unlike the microfluidics project where Devi and Pasha went back into the lab to account for anomalous data, in the Hodgkin-Huxley project, they tried to understand the anomaly in its own right—as "interesting data." In regard to the class goal of teaching students how to use data, the Hodgkin-Huxley project shows Devi gained more confidence as a researcher as she took on an investigative role in pursuing interesting data and then sharing that investigative logic with reviewers.

For the Hodgkin-Huxley project, Devi and Pasha decided to study the velocity of the propagation of an action potential down the squid axon. They hypothesized that the action potential would decrease when the membrane capacitance was increased. The data confirmed their initial hypothesis, but some of the data were anomalous beyond a certain distance. Devi explained: "We [looked at] the data, and what we found really interesting was when the membrane capacity was above 5.6 microns per centimeters squared, there was an action potential. But when we increased it to 5.61 there was not an action potential. So, there's this .01 difference that [resulted in an] action potential or no action potential." Since the theoretical model did not explain the disappearance of the action potential, Devi and Pasha collected additional data at the 5.7, 5.8 microns, and higher values. Upon investigating this phenomenon more closely, they proposed a reason for the change in the action potential behavior: if the sodium and potassium current balanced each other out, then the nerve would not fire and produce an action potential.

Armed with their new insights into the behavior of nerves, Devi and Pasha started to work on the storyboard for their oral presentation. In designing the story of their research, they wanted to convey not just what they found but how they deduced this finding. This was a major rhetorical shift for Devi and Pasha. In the microfluidics experiment, they were mainly interested in providing their readers meaningful interpretations of the data; here, they wanted to share their thinking process with their readers. Devi explained this process: "[Pasha] and I had a lot of graphs. And then we looked at the graph that Pasha plotted out with the capacitance and velocity and as we looked at it, we decided that was our main hypothesis—to show the relationship between velocity and capacitance. And when we looked at it, we saw that there's no action potential being generated in this area. Then we were like, OK, that's good, that's actually an interesting finding to talk about."

In her explanation of the drafting process for the oral presentation, Devi described several concepts from the storyboard lecture, with the first goal being to understand what data you have and then figure out why those data are interesting. Focusing on interesting data takes some confidence on the part of the researcher; she must be confident that she understands what is going on so that she can identify what is interesting in the findings.

In presenting the results from their research, Devi and Pasha led their narrative with what they considered their key finding: the disappearance of the action potential after 5.61 microns. The plot they chose as evidence of that finding was visually simple, with two region zones imposed over the data to enhance the division between action potential and no action potential areas (figure 4.3). A supporting equation provided theo-

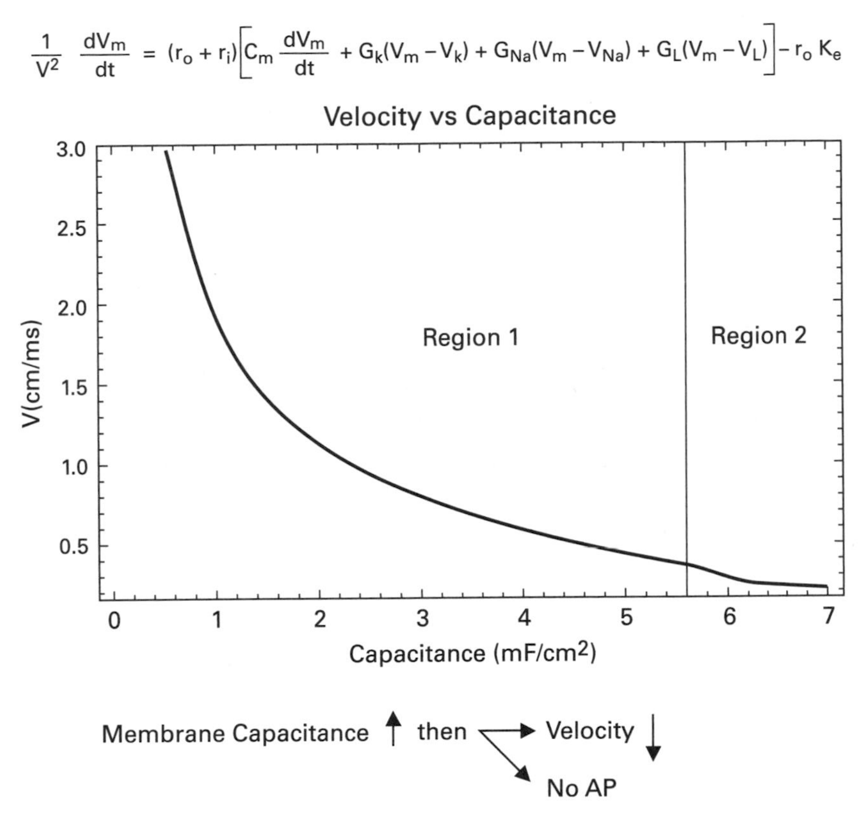

Figure 4.3
Results figure from Devi and Pasha's Hodgkin-Huxley presentation, showing their key finding

retical support for the plot, and the supporting text summarized the message for the audience.

This figure was followed by supporting evidence of the microscopic mechanics of the channels to explain the disappearance of the action potential. The supporting slides provided further evidence of why there was no action potential (figure 4.4a–c). For this part of their presentation, Devi and Pasha added their own graphic, a table, to represent six regions that showed the opening and closing of ion gates with the flow of sodium and potassium through those gates. For their final evidence, they included a slide with a plot, showing how an action potential can be generated. A summary slide concluded the draft presentation.

a) Why no Action Potential?

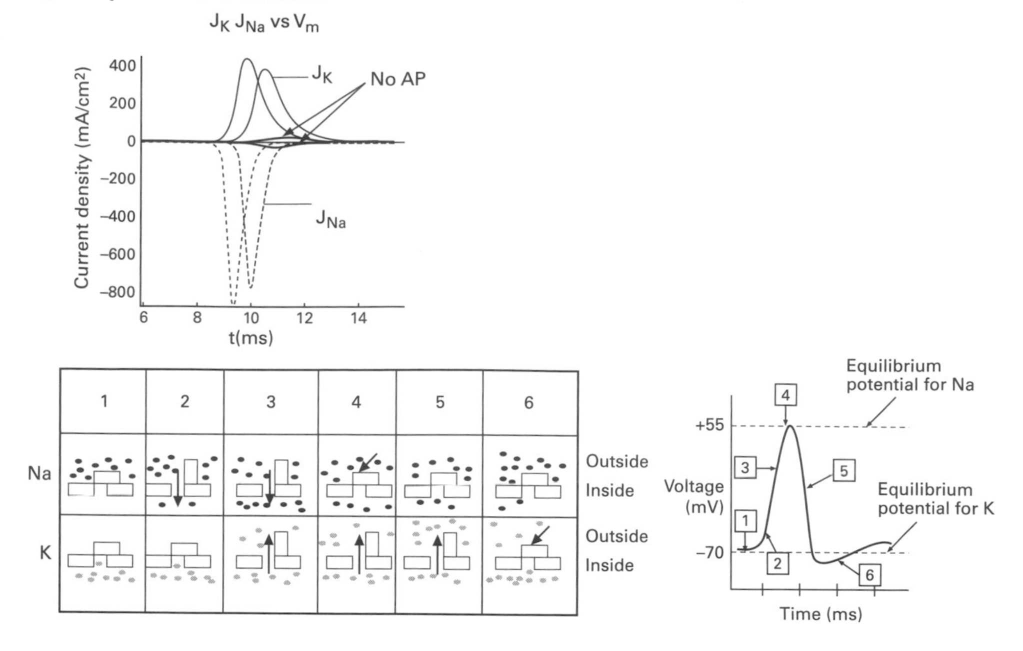

b) Influence of J_{Na} on the Propagation of the Action Potential

Membrane Capacitance (mF/cm^2)	G_{Na} Value where AP occurs
5.7	122
5.8	123
5.9	125
6.0	126
6.1	128

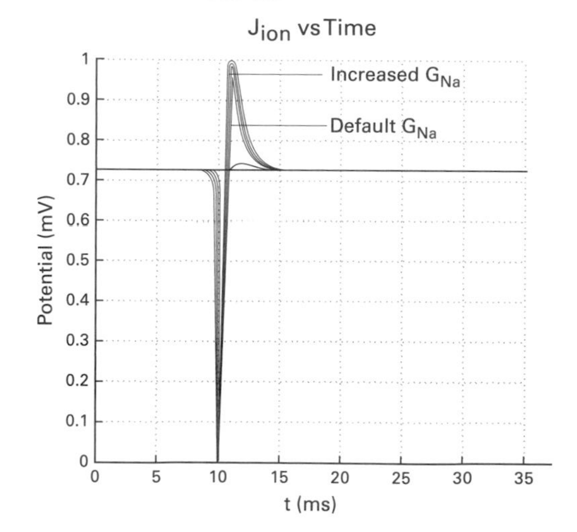

c) Summary

Increasing the membrane capacitance decreases the velocity of the propagated action potential

Increasing the membrane capacitance above a critical value impedes the propagation of the Action Potential

Action Potential fails to occur at this critical value of the membrane capacitance because Jion is not adequate enough to increase the membrane potential above the threshold value

Figure 4.4

Series of supporting results figures from Devi and Pasha's Hodgkin-Huxley presentation

Joel Voldman, the faculty member who critiqued Devi and Pasha's practice presentation, focused on the plots in his comments to the students: "I overall really like your analysis of why increasing Cm kills the AP and I do not think you should make any major changes to your scope. Instead, I would spend more time on the analysis section (figure 4a–b), adding plots/slide to explain your model not only diagramatically (figure 4.a) but also quantitatively with reference to JNa and JK." In addition to his comments about ways to improve the technical presentation of the data and more development of the analysis, Voldman suggested some stylistic changes:

Go through and edit the plots to make them more readable. For instance, make the axis labels and other fonts on the plots big enough to read, remove extraneous dotted lines, increase linewidth, make the colors stand apart from each other better, etc.

Focus the plots on the regions of interest. For instance, you do not need to plot all of time, and you may be able to focus in on the element that you want the reader to see, which is Jc and Jm at high Cm. You might even stage the plots to make it easier for the reader to see how the currents change as Cm increases.

In his comments, Voldman did not quibble with Devi and Pasha's results. Instead, he encouraged them to focus on expanding their analysis, adding more visual evidence in support of their conclusions. He also suggested that they attend to the aesthetic aspects of the plots that improve meaning: axes, regions of interest, and font size. He noted that these elements improve the readability of the plots.

Devi and Pasha took up many of Voldman's suggestions, internalizing that "even though we understood how that happened, the audience does not understand how that happened." In the final presentation, they kept the key results slide (figure 4.3) but provided additional slides in support of the main finding. These three additional slides expanded on their reasoning of why the action potential disappeared. They also focused these plots on the regions of interest, as Voldman suggested. However, they ignored his advice about making the plots more readable, which Voldman noted in his comments on the final presentation: "This was a much-improved final presentation. There were some really nice points, most notably the experiment at the end where you compensated for the lack of action potential by increasing GNa. That was really nice. That said, there were some issues that could have been improved. The plots were still hard to see and a bit fuzzy, and could have benefited from some staging."

Voldman praised Devi and Pasha for adding to their analysis of the data, but he again criticized them for not attending to readability issue (the plots were "a bit fuzzy"). Such comments were meant to reinforce the second concept in storyboarding: design figures that make the point that you want to make. One way to help reinforce the point of a slide is to use animation or staging to show a logical sequencing of events on a single slide.

Ultimately Devi left Quantitative Physiology with a deeper sense of the desires of scientific audiences: "The audience wants to hear that story. They do not want to hear yeah, we did this experiment and [such a] success. They'll be like [great for you] but how does that teach me anything?"

On one hand, by internalizing that a scientific audience wants an interesting and plausible story more than the correct theoretical fit to data, Devi could think about the computer simulation data in ways that permitted her to be more exploratory with the data. On the other hand, some of Devi's success this time in presenting her data may have come from the genre in which she was asked to perform. Oral presentations encourage narrative storytelling more so than written scientific genres do, especially for students. Indeed, when we asked Devi to reflect on storyboarding, she noted that the method seemed much better suited to oral presentations: "When the presentation came we had to make sure that [we got] the point across in a way that an audience could understand. So the first thing was to explain to them how we understood it: What did we do first? We did these experiments, and we looked at the raw data. Then what did we do next? Then we looked at why there's an interesting feature in our data. And then how did you process that? How did we explain it? So that was our story."

At the end of the semester, Devi noted that research and writing "are an on-going process. Much of the work is done at the beginning [of the research process] as well as at the end." From this case study, we see how faculty feedback from their perspective as professional scientists pushed Devi to engage with her data in ways that simply telling her about aesthetic dimensions of plots would not. The plots themselves were central to the conversations between Devi and her reviewers, becoming the locus of activity in their talk about the interpretations to be made of the raw data. And finally, we see how an authentic research experience, along with an iterative feedback and revision cycle focused on data, can help students learn the technical content of a course and improve as scientific writers and speakers.

Case Study 2: Julia—"Not Giving Up" on Noisy Data

Julia, a biological engineering senior, came to Quantitative Physiology with much more research experience than Devi. At the time Julia enrolled in Quantitative Physiology, she had already taken Laboratory Fundamentals in Biological Engineering, the class profiled in chapter 2 of this book and had also done several undergraduate research projects with MIT faculty. Those activities had equipped her to start thinking about scientific research with an emerging professional identity, and Julia's survey answers at the beginning of the semester showed her familiarity with the role of argument in scientific writing. When asked about the purpose of the scientific article, she wrote: "[A scientific article should] present and analyze a significant finding in an honest, well-informed manner. The data should be presented clearly without being overshadowed by qualifying words like 'clearly shows' or 'immediately suggests' that lead the reader. The data should speak for itself (although analysis is also important). All procedures should be replicable by other labs." In her survey answers, Julia noted that certain words can diminish the persuasiveness of an argument and obstruct a reader's understanding of data. In her comments, we also see her underscoring the importance of veracity in research, and one way to establish the veracity of a researcher's claims is through repeatable methods.

In terms of the two class projects produced in Quantitative Physiology, Julia ultimately produced a well-received research article and a well-received oral presentation. She received a high grade for the course, but she often seemed frustrated by the "student-ness" of the lab activities and wanted to move beyond the constraints of the lab to work that more closely simulated professional activity. Despite her restlessness, Julia did develop a deeper sense of the rhetorical aspects to the presentation of data through the two research projects.

For her microfluidics project, Julia and her partner, John, investigated the osmostic response of mouse cells in three different substances. While Devi looked at the cell responses under varying concentrations of the same substance, Julia investigated the cell responses to various substances: phosphate-based solution, urea, and D-glucose. In terms of making sense of the data, Julia came to the project with a good understanding of biology and how to organize the raw data, making it possible for her to start by looking for interesting findings in the raw data. Indeed, she would obtain interesting data in the first experiment.

During the experiment, the cells responded more quickly than expected, which meant that Julia was able to gather just a few data points between the beginning and the end of the cellular response. As she explained, the limited data points outlined the trends but did not "provide enough points in order to completely convince me." Julia contextualized the problem as "noisy data." In attempting to reconcile the problem with her experimental data, she reflected on her other research experiences where she had been taught to work with noisy data rather than "giving up" on the data: "I need to make sense of this data and I'm not just going to just say things randomly.... I wanted to be able to sit down with the data and try to make sense of it as it is ... without actually changing it."

One of the strategies that Julia had learned for dealing with noisy data was normalizing the data, a way to extract information out of the data without actually changing them in a way that scientific readers would perceive as manipulating the data. Julia's approach suggested an awareness that raw data can tell many stories, and it is up to the researcher to decide how to extract a meaningful story out of the data. "In normalizing [data]," she said, "you're keeping the same ratio but you make it in a way that people can understand. . . . I want to, at least, present a story here. So I'm not just going to leave the graph as it is; I'm going to try to work with the data and that's sort of my attitude towards it. Once we have the data, there's something in there. I do not like giving up on the data that easily."

Julia's concerns were ultimately reflected in her choice of visuals for her research article. Julia and John provided plots of both raw and normalized data (figure 4.5). Julia explained her justification for the raw data plot (figure 4.5a): "the reader deserves to know the fidelity of your data, like what was actually measured." In presenting raw data as simply raw data, the author makes no claims about those data other than that he or she has enough confidence in the data to show them to readers. Julia saw her scientific ethos withstanding the criticism of reviewers: "We are presenting some sample raw data before we filtered through everything just so people know what we did to the data and if we did something invalid they can catch us on it and I think that they should have that right."

Julia's second plot of normalized data (figure 4.5b) represented her attempt to make sense of the noisy raw data. After normalizing the data, readers could compare the differences in time and rates of change. Without normalizing the data, Julia explained, "It's really hard for your eye to line them up."

Taken together, Julia saw these two plots making a logical progression in the results that allowed readers to move from the raw data, to the normalized data, to the statistical evidence. Julia had, in essence, demonstrated the third concept in storyboarding: the illustrations should make a logical sequence from one figure to the next.

Joel Voldman, the faculty member who responded to Julia's draft, liked a lot of what she had done. He thought the images were well chosen, especially the histogram equalization, which removed "the human subjectiveness from the image analysis." Voldman did suggest that Julia and John include the standard deviations on the plots "so the reader can more easily judge the variations in the measurements." He also asked that the graphs be replotted without the connecting lines, "as those lines assume a linear variation between timepoints, which may or may not be true." Finally, he commented on the "fit" data: "It would be helpful to the reader to plot the fitted expo-

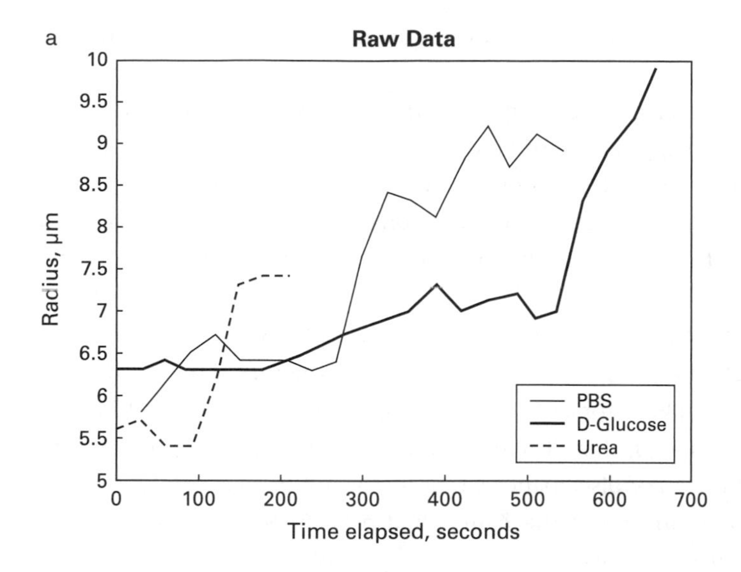

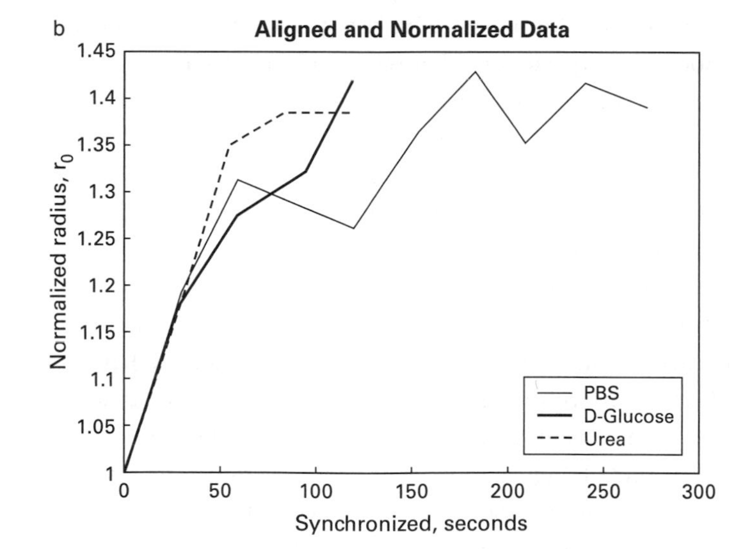

Figure 4.5
Results figure from Julia and John's draft microfluidics report

nential along with the data, so that the reader can qualitatively judge the goodness of the fit."

Voldman's comments reflect his readership as a professional scientist. He was suspect of trend lines that gave the appearance of "too good" data, hence his desire for standard deviations noted in the plots. He approved of visual representations that appeared to remove human subjectivity in methodological choices and of the interpretation of the data, noting that with a "few adjustments," the "team will have a convincing story."

In the final revision, Julia and John incorporated Voldman's feedback, adding error bars to the raw data plot and fitting the normalized data exponentially (figure 4.6).

Voldman praised Julia and John for their revisions, especially their "outstanding data analysis." He cautioned, though, that "the team does try to read a bit too much into the data, trying to find causes of discrepancies that might not be real." He also noted that even though the data in one of the figures had been fitted exponentially, the figure would be better with error bars.

In Voldman's comments to Julia and John, we see his rising expectations for the pair as he continued to drive home the expectations that scientific readers have. His comment about reading "too much" into the data was cautionary. Julia's desire to work with noisy data showed a level of maturity not possessed by other students in the class, but it can also be a slippery challenge, and Voldman notes that overanalysis of data is one of the main complaints of scientific readers. The pitfall of overinterpreting the data would become an important theme for Julia in the second project.

For her second project, Julia and her new partner, Pia, chose to investigate the phase after the action potential (when a nerve fires) called the "absolute" and "relative" refractory periods. They wanted to know how certain variables (K_h and K_n) in the Hodgkin-Huxley model affect the nerve's time to recover from the stimulus that created the action potential. As Julia explained in her presentation, lengthened or shortened refractory periods are an underlying cause in many neurodegenerative diseases. When Julia was asked why she chose this particular project, she explained that she wanted a project that was not "trivial" but was also "well defined": "We didn't want to do one of those projects which was really open-ended, where you wouldn't really know how to test it necessarily or you would feel that your results were not significant."

As in the microfluidics project, Julia desired meaningful results that were grounded in authentic data analysis procedures. One of her strategies was to gather a lot of data. In the microfluidics experiment, she had been disappointed that she and her partner had not collected enough data points. So for the Hodgkin-Huxley project, she wrote a

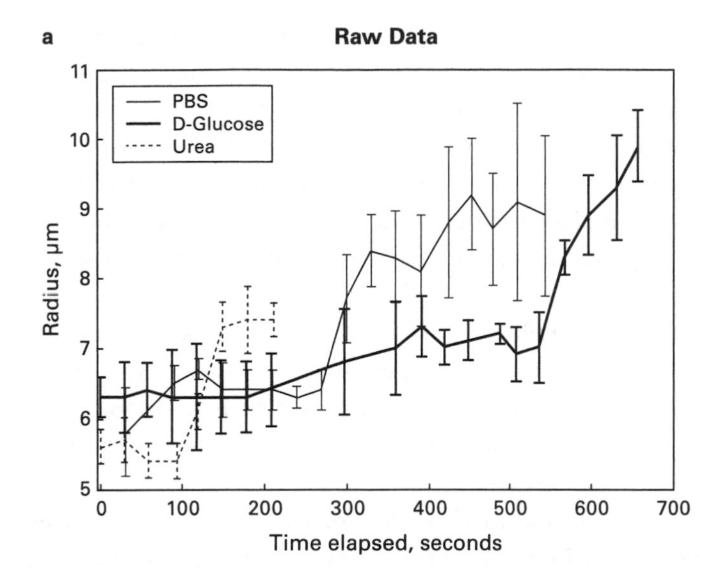

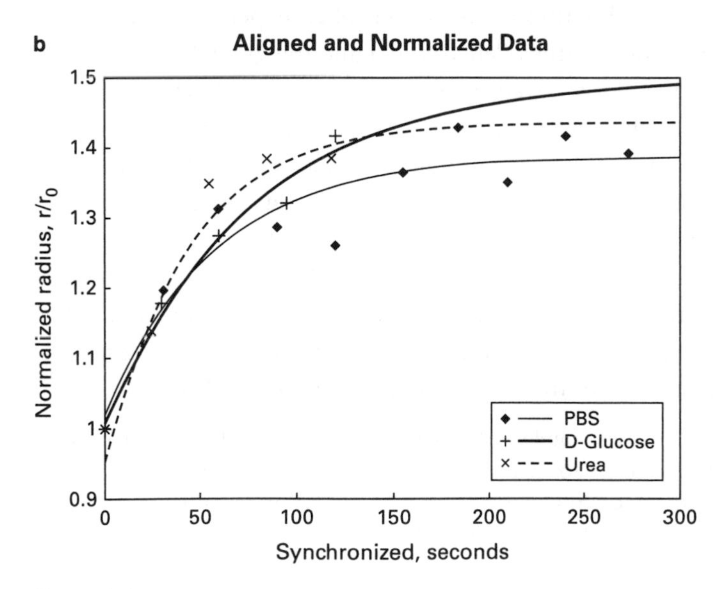

Figure 4.6
Results figure from Julia and John's final revised microfluidics report

computer algorithm to run a multitude of computer simulations. Besides, as Julia explained, she did not "have the patience to actually click every number." Her program yielded 16,000 data files that then needed to be analyzed. She "poked around and figured out where the data [were that we were] interested in." Then she wrote another computer script to analyze those data. In looking at the data, Julia noticed that sometimes the computer algorithm returned anomalous results. She noted that this was one of the issues with computer simulations; they are imperfect models of the natural world.

The limitations of the computer simulation data became apparent as Julia and Pia drafted their presentation. Julia explained that when she and Pia "drew up some things," they realized that some of the data did not support their hypothesis. Again, Julia was concerned about the data. Data were a reflection of the author's ethos, so they have to be plausible as well as interesting.

As Julia explained her thinking behind the slide design, she drew on her other research experiences. One particularly formative experience was with a graduate student mentor who had taught her to think through the results of an experiment: "He would say, what is the graph I want out of this? Can you imagine the graph? Imagine what trend you want to see.... How do you want your audience to react? What is the eye seeing? Your eye isn't seeing what the audience is seeing." Through the graduate student mentor, Julia had learned to "see" her raw data translated into visual evidence from the perspective of the audience. Being able to present data effectively is about eliciting a response from the audience—one that says, "Yes, I believe you."

This attention to the "eye" and "seeing" was also evident in the design choices that Julia made for her presentation slides. From the use of colors to the scaling and placement of images on the slide, Julia and Pia designed their presentation for a scientific audience. Julia talked at length about those design choices and their effect:

> So what we ended up with was this very simple, just white background and what I intended to do was use large fonts and a lot of color.... So we picked out a color scheme, like orange is going to be K_H and pink is going to be K_N, and action potential is green, no action potential is red, just so that people have something to ground themselves in. At the same time, if people are color blind or if you have a black and white printer, we had to make sure that the graphs had the appropriate differences in markers just in case.... I found that particular uses of color were very effective: particular uses were not, like when people took technical pictures and put them up and you couldn't see anything because the yellows and the light purples all disappeared. I was like, "Okay, those colors are out, never to be used again."

Julia's observations are reminiscent of the kinds of advice usually found in technical writing handbooks about aesthetic choices that contribute to readability. Julia and Pia included such ideas in their presentation slides. The methods and results slides used repeated colors to link concepts across slides. Such features are important to readability but their selection comes late in the research writing process, after the data have been selected and a story assigned.

Like Devi, Julia and Pia foregrounded their key result in their presentation with a single slide—a plot that showed the difference in the relative and absolute refractory periods (figure 4.7a). In the following supporting slides, Julia and Pia showed "interesting" elements of the data or anomalies that needed further explanation (figures 4.7b to 4.7d).

Voldman applauded Julia and Pia for the "impressive" quality of their practice presentation and offered only minor comments:

> You should consider plotting the data in [the first figure] normalized to the $K = 1$ value rather than the absolute values. As plotted, the effect of the K's appears minor because you are plotting on an absolute y-scale. What you really care about is not whether the refractory period is 9.2 or 7.9 ms, but that it has decreased by 15%.
>
> On [the last figure], I think that the plot of refractory period vs. temp does not enforce your message. What you'd like to show is a plot that has the abs/rel refractory period at 6.3 C, and then show that increasing K can be compensated by decreasing T.

As in his comments on the microfluidics project, Voldman offered Julia advice on the presentation of her data but did not criticize the data themselves. Here he did not ask Julia to extend her analysis of the data but to reconsider if the visuals supported her spoken message. He reinforced his comments by suggesting an alternative narrative to what was shown in the plots.

Julia and Pia took Voldman's advice in revising their presentation and normalized the data in their main results slide (figure 4.8a) and added plots of the absolute versus relative refractory periods in a supporting slide (figure 4.8d). The other two results figures remained unchanged (figures 4.8b and 4.8c). Voldmann praised these changes: "This was a most impressive study and presentation. I was really impressed with the depth into which you analyzed the effect of K_h and K_n on refractoriness, from the methods to the changing of K_t. The slides were clear and convincing, and you were able to effectively communicate a large amount of information in the time allotted. Great job." In his comments, Voldman praised Julia for the depth of the analysis and the presentation of that research. He linked the "clear and convincing" presentation of the technical details with the communication of those findings, and he applauded Julia and Pia for being able to present "a large amount of information" in the scant ten minutes allotted for the presentation.

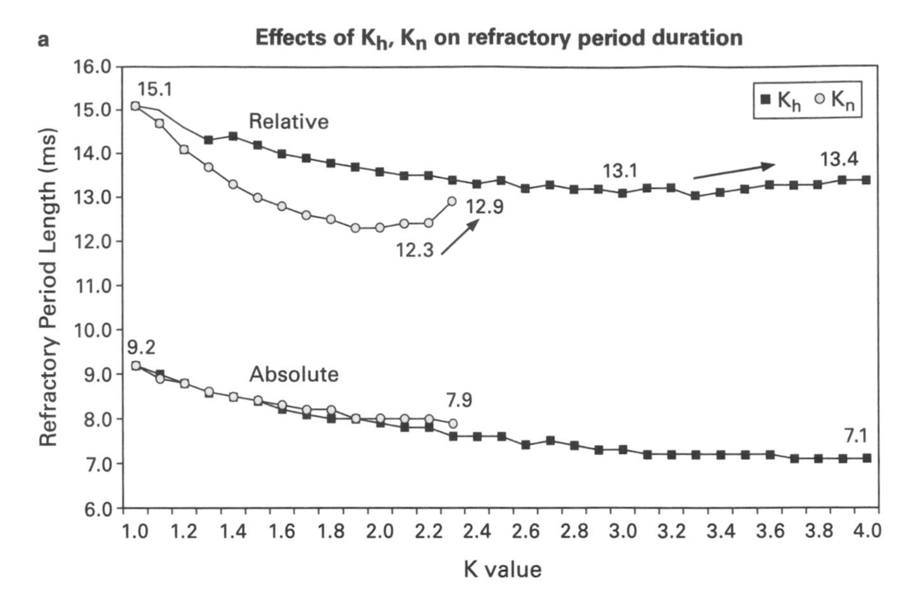

- Turnaround in relative period duration for high K
- Relative more sensitive to K_n; Absolute equally sensitive to both
- No action potential generated for $K_h > 4.0$ or $K_n > 2.3$

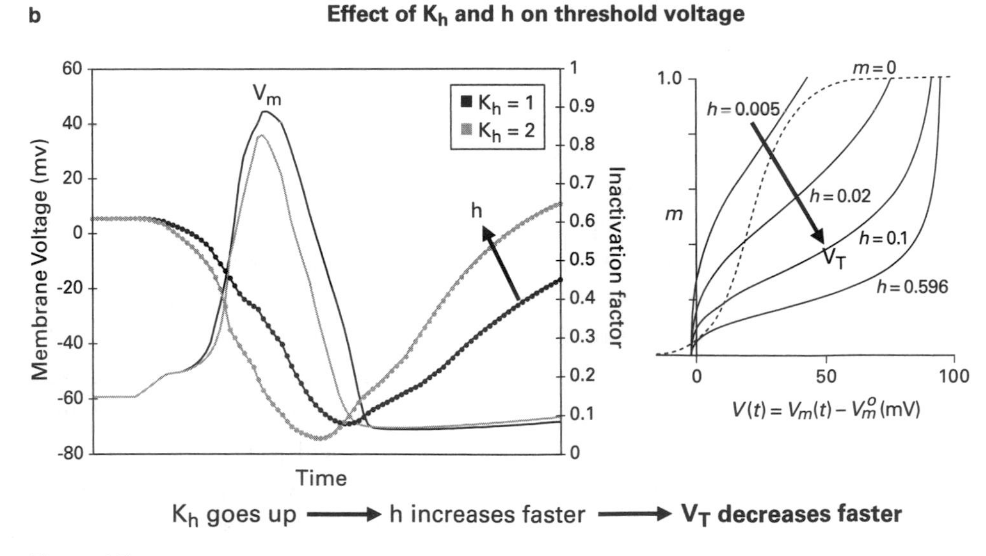

Figure 4.7

Results figures from Julia and Pia's draft Hodgkin-Huxley presentation. Figure 4.7a shows their key finding. Figures 4.7b–d are supporting figures.

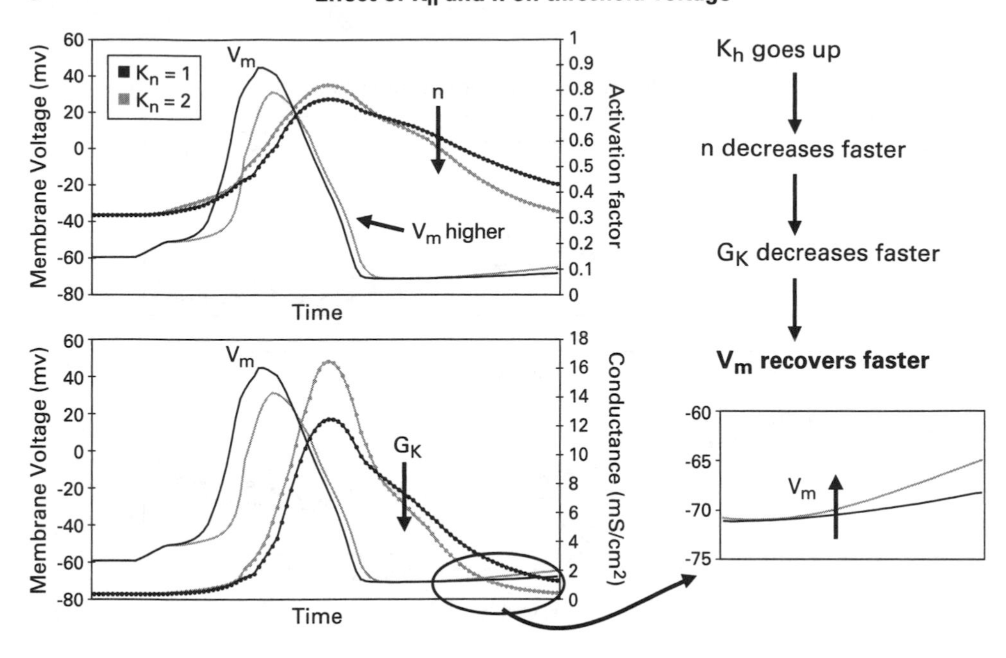

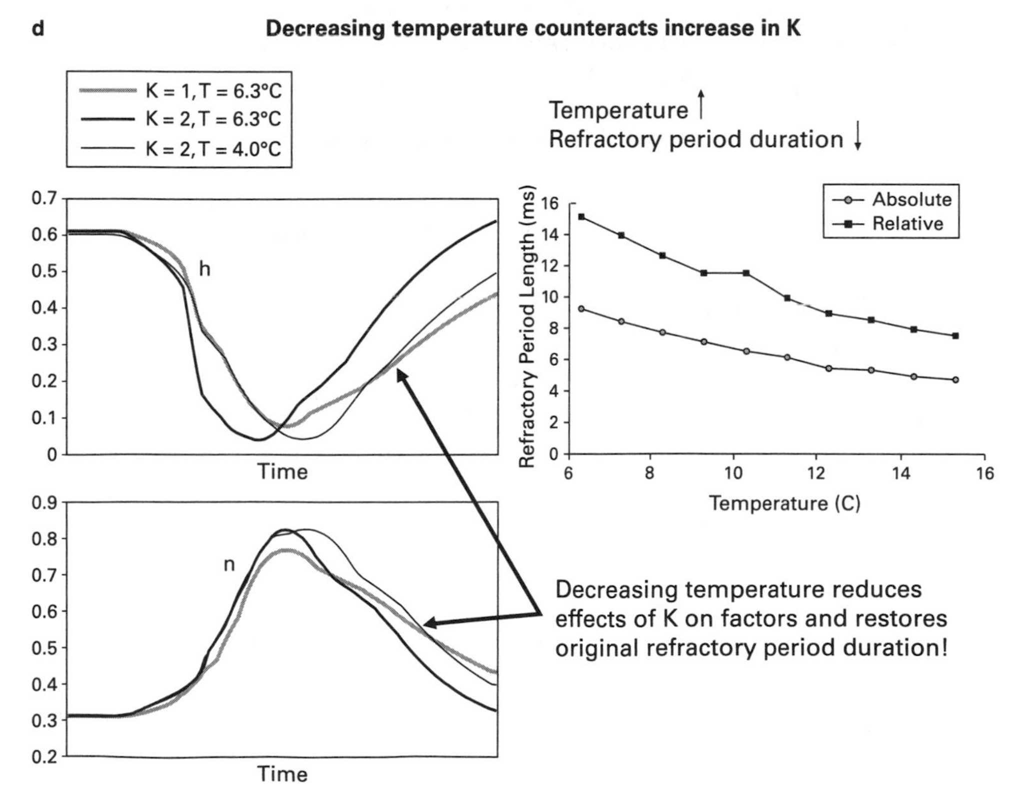

Figure 4.7
(continued)

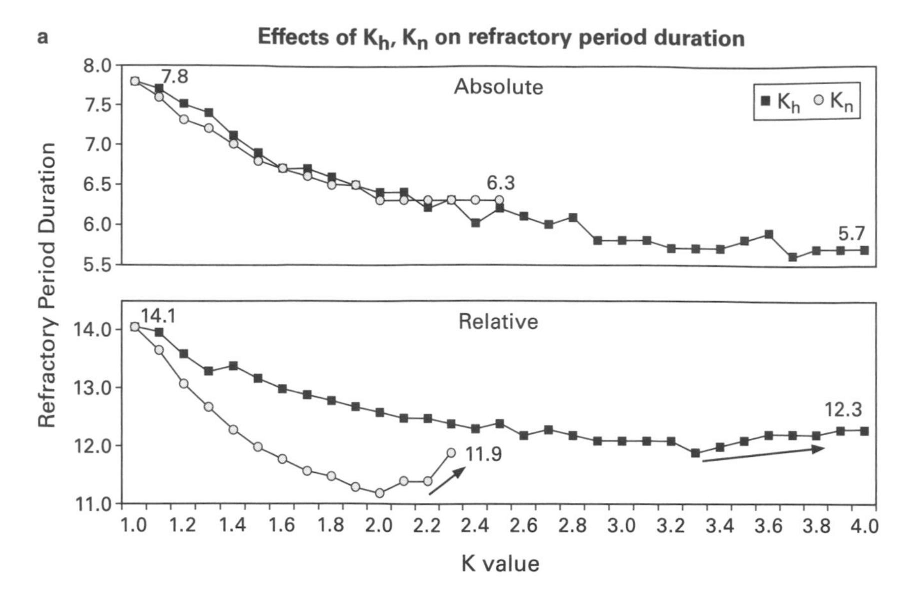

- Turnaround in relative period duration for high K
- Relative more sensitive to K_n; Absolute equally sensitive to both
- No action potential generated for $K_h > 4.0$ or $K_n > 2.3$

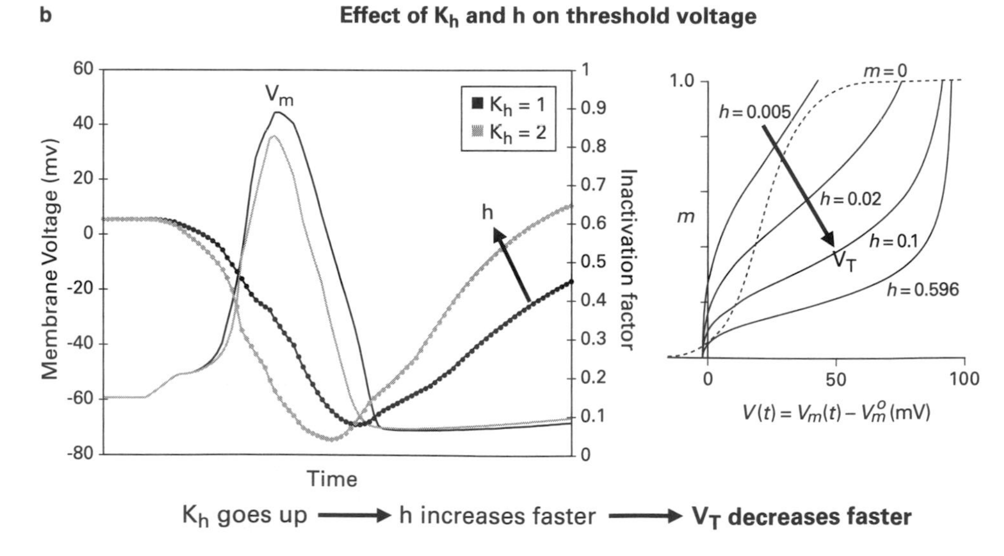

Figure 4.8

Results figures from Julia and Pia's final Hodgkin-Huxley presentation

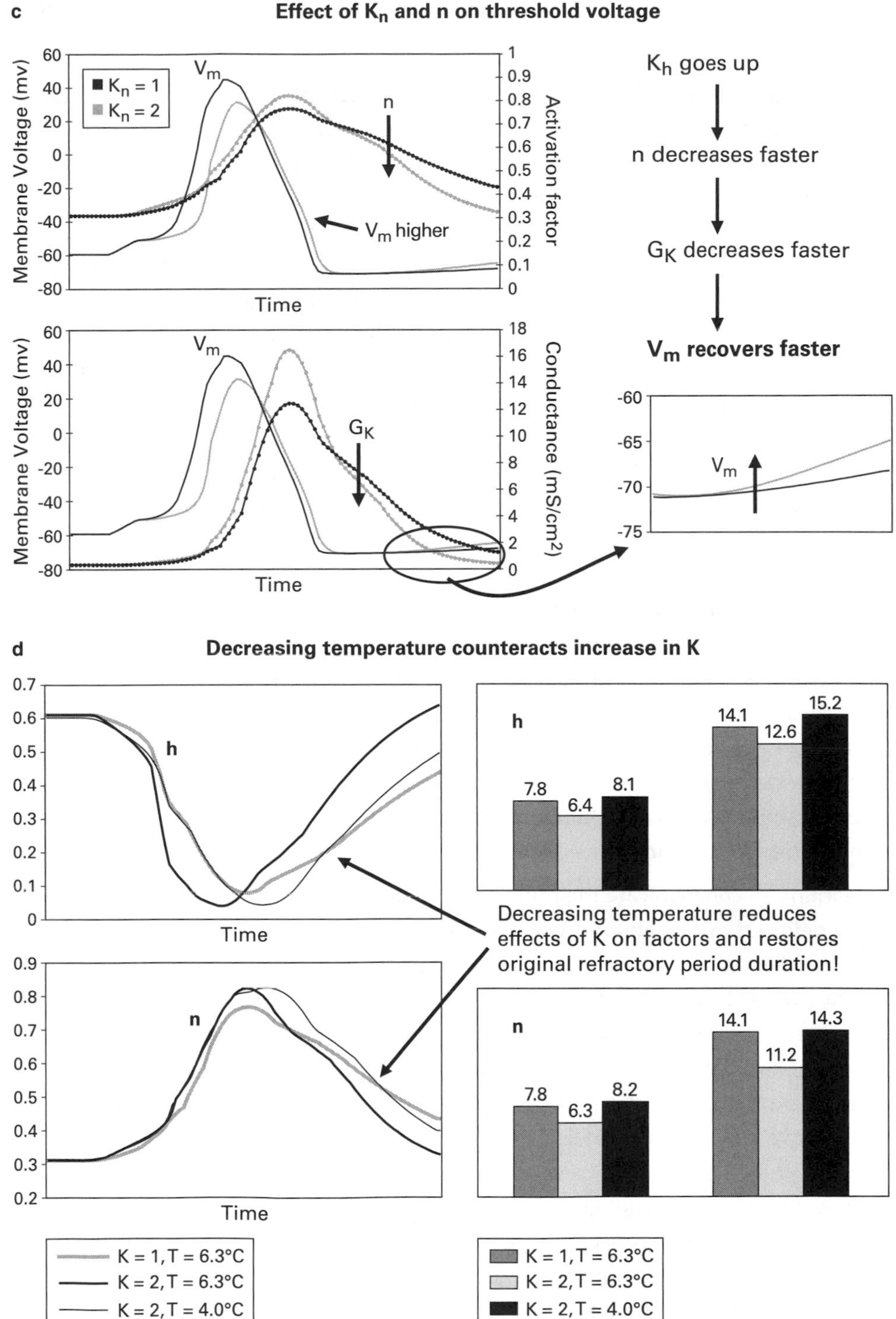

Figure 4.8
(continued)

By the end of the semester, Julia's ideas about scientific communication had not changed much, but they had grown deeper: "I've always valued [the visual display of scientific data] a lot but this class reinforced my specific strategies for presenting graphs and tables clearly without distraction/confusion."

In her interviews and projects, we see Julia not so much struggling with the basics of making sense of data but how to make the most of data without overreaching her analysis. Finding "interesting" data was not simply a task of pursuing an anomaly in the data. For Julia, what was interesting about data was working with less-than-perfect data (i.e., "real data") and making sense of them "as they are." We also see her developing a deeper understanding of the aesthetics of visual displays of data—how visual choices help an audience see what the researcher sees in the data and reach the same conclusion. Finally, like Devi, we see Julia's sense of telling a story with her data crystallizing in the Hodgkin-Huxley presentation, "where it was important to convince the audience of [her] vision."

Summary of Quantitative Physiology

The case studies of Devi and Julia show how students learn to make arguments with data, their evolving understanding of how to manage data, and the standards for using data as persuasive evidence. Their case studies also show a portrait of how wrestling with data can become a formative part of students' learning technical communication. The case studies of Devi and Julia along with the survey data from the class suggest certain findings for teaching students how to write about scientific data:

In learning to communicate their findings, students first talk about their work descriptively before they talk about it persuasively. Students follow a developmental trajectory that begins with talking about what they are doing methodologically and how they are doing it to why certain data are interesting and what the limitations or possibilities are with those data. In making this transition, students come to distinguish between the raw data gathered in the lab and the visual evidence in research articles. By focusing on how to translate raw data into a compelling story of research findings, students move beyond static understandings of data to dynamic interactions with their research processes. Students come to understand that argument is not woven around data; rather, it comes through the process of collection and analysis. Every methodological choice in data collection and analysis is a decision that allows researchers to foreground certain results and not others. From the study of the students in Quantitative Physiology, it appears that learning the aesthetic choices that help an audience understand what the researcher "sees" comes last in this developmental process.

Different communicative genres lead to different learning experiences in understanding how to use scientific data. In the microfluidics lab, students wrestled with the practical considerations of bench work: cells die, equipment breaks, and biological systems rarely function according to theory. In the Hodgkin-Huxley project, students wrestled with the dilemmas of computer simulations. It is easy to plug in numbers in a computer simulation without understanding the actual biological system it is supposed to model. Devi summarized the differences between the two projects as follows:

> In the microfluidic you can get a whole lot of data that does not make any sense and it's hard to understand if it's because you did something wrong in the experiment or it's an interesting area that would be worth looking into.... But in the Hodgkin Huxley model I think it's a big question and we take big questions for granted. We believe that if you plug into these values you should get these kinds of data out of [the computer simulation program]. It's an equation; how can equations go wrong, right?

In the end, each project yielded quite different data that challenged students' ability to work with data in different ways. Offering students multiple opportunities and challenges in working with data seemed to facilitate deeper understanding of how to make arguments with data.

Although the projects yielded different kinds of data, both taught students the importance of seeing data and communicating that vision to the audience. A researcher goes into a project with a clear purpose and a good sense of what is expected but not necessarily what is going to be interesting about the results. It is only in the process of gathering raw data and translating those data into some written inscription that what is interesting becomes apparent. Even the same data set can yield quite different results, thus depending on the way you approach that data, you can get a whole different story out of that same data. Selecting the interesting data from the outputs, the researcher returns repeatedly to the data for further analysis in finally making his or her final case with the visual evidence. In the case of Devi, being able to "see" the data meant returning to the lab and revising her analysis. Julia found what was interesting early on but continued to refine her analysis until she felt the audience could also see her vision of the data. An important turn here for both participants was coming to see audience not as a static group of individuals onto which ideas are cast but as individuals who are persuaded by ideas using the representational system of a particular scientific community.

Faculty feedback that modeled authentic, professional feedback helped guide students in developing their understanding of using data in making scientific arguments. When quantitative physiology faculty responded to student writing, they responded often in ways that modeled professional review. Faculty Bhatia and Voldman remarked on how data were presented, if that presentation supported certain claims, and when conclusions were too ambitious to draw from the data presented. Such feedback can be thought of as "authentic feedback" because it uses the terminology and tone of professional scientists in the evaluation of scientific communication.

Yet this finding does not mean that faculty completely adopted a professional identity in responding to student writing and presentations. At times, their responses took a teacherly tone, praising students for good work or reminding them to attend to the details, such as labeling graph axes.

In engaging in authentic feedback, faulty drew on their deep disciplinary expertise, but they were also guided by the assignment scoring rubrics that included criteria for storyboarding as well as technical clarity and insightfulness. The scoring rubrics helped remind faculty that their own role was to offer high-level comments, not mark calculation errors or offer generic feedback.

Most important, students responded to this feedback. Authentic feedback challenged students to return to their drafts with a sense of themselves as professional researchers—which they did and returned with revised manuscripts that typically went beyond superficial editing. In doing so, students began to adopt the language of argument and evidence in the use of data in talking about their research.

5 Writing and Speaking Collaboratively

The development of professional identities often takes place for engineering students when they undertake complex design and experimental projects as part of a team. However, their technical design and research expertise, although it is central, may not be the only criteria for professional success. The ability to write and speak clearly about design and research is critical; moreover, these specific tasks serve as a student's introduction to his or her professional discourse community, a key step in developing an identity as a professional scientist or engineer. Yet if learning to write and speak as a professional and make arguments from data were not challenging enough, engineering and science students must also learn to do this collaboratively. As they master the collaborative communication task, they must develop new writing processes and use relatively unpracticed team skills to complete their documents or their presentations. Interviews and surveys in this chapter were focused on students engaged in this process and asked:

- How do students efficiently and effectively learn to write and present collaboratively?
- What specific team skills are central to this task?
- How are those team skills best learned? By instruction? By mentoring? By experience?

Collaborative writing and presentation are common in the engineering and science professions. In fact, nearly 90 percent of engineering professionals report collaborative writing and speaking as an important part of their jobs (Ede and Lunsford 1983). Other researchers have documented the prevalence of collaborative writing in the professions (Anderson 1985). Faigley and Miller's survey described multiple authorship as a "major difference between writing on the job and writing at school" (1982, p. 567). Although collaboration varies by context, "co-authorship [was found to be] especially common in professional and technical occupations" (Faigley and Miller 1982, p. 567). Schulz and Ludlow agree that the "development of group writing skills is important for engineers to be successful in the workplace" (1996, p. 227). Louth argues that collaborative writing experience is essential (1989, p. 3).

While the importance of collaborative communication is readily acknowledged, the literature does not easily converge on a definition of this activity (Speck, Johnson, and Heaton 1999). The descriptions of collaborative communication range from projects in which all aspects—research, invention, composition, editing, and revision—are done collaboratively to projects in which individual subject matter experts write while a project manager assembles and edits. Some authors argue that all writing is collaborative and that the construct of individual authorship no long holds true, if it ever did (Ede and Lunsford 2001). A precise definition of collaborative communication is challenging to determine; what seems more to the point is thinking of collaborative communication as a set of practices that professionals implement in their specific disciplinary contexts in response to specific writing or presentation criteria. What characterizes professionals is that they know the criteria for successful communication in their discipline and have some experience in how to achieve those criteria collaboratively. Undergraduate students usually are unfamiliar with this process.

Anson and Forsberg describe the process of students' learning about authentic communication processes as a "transition that writers make as they move into new and unfamiliar writing contexts" (1990, p. 204). It is the process of learning to accommodate and negotiate the new contexts that helps students learn to communicate in their professional setting. Winsor comments that this process also involves "achieving an appropriate role in an organization" (1996, p. 10). Miller describes technical or scientific writing as "an understanding of how to belong to a community" (1979, p. 617). Paradis and Dobrin note that writing has a social function within a group context, establishing status and authority of the writer and maintaining group dynamics (1985, pp. 292–293).

Collaborative communication, then, might be best understood as the dynamic achievement of a shared communication product through a process within a group in which the writer assumes a functional role. The more interesting question is not, "What is collaboration?" but instead, "How does a novice writer [or speaker] assume a role in this collaborative effort? What helps him or her do that?"

In our classrooms and despite our efforts to design authentic communication tasks, the process of collaborating on writing or speaking is often not clearly specified, and students are sometimes left to their own devices. The pedagogy that could support that process is not always well understood or always practiced in a consistent and timely way. The resulting confusion impedes students' ability to learn the professional communication practices of their disciplines.

Students at MIT usually know how to write as individuals. Most of them learned an individual writing process beginning in their elementary school years, and when we interviewed them, most could clearly describe their successful processes. As we expected from these students, their individual writing processes are thorough, creative, and individualistic. And since they are accustomed to writing alone, their individual writing is free from interpersonal conflict. Students rarely report using external reviewers, although they commonly edit and revise their individual work in some way. Most of the time, their audience is their professor. Feedback on that writing may come from peer editing, but more often the writing is commented on and then graded by the professor. Students describe varying individual writing processes depending on the format or the medium. Written deliverables are usually more scrutinized, they report, while oral presentations can be quickly composed the night before. But in either case, the choices made about individual writing process are those of the individual student.

Once students are in a communication-intensive, team-based engineering project, they are plunged into a very different writing and composition process. They must communicate with their professors and mentors and project sponsors, and they must also communicate with their team members since many complex engineering projects are divided into subsystem teams. Writing is no longer a private and individual act; prepared or not, they have joined a community of sorts.

Bruffee (1984) describes collaborative writing as a public conversation in which individual thought is made external, reflected on, reinternalized, and reexpressed, thus becoming knowledge that is shaped by participation in a community. As students participate in this process—clumsily and often with resistance at first—they begin to grasp "how knowledge is established and maintained in the 'normal discourse' of communities of knowledgeable peers" (p. 646).

In this more public conversation, students discover that collaborative communication process is something more than individual composition. The process requires careful project management and some consensus among team members. Communicating as part of a team means less privacy and less individual control. Students share praise for their work, but sometimes they also share criticism. Moreover, a critique that comes for what a student perceives as the failure of his or her teammates can provoke anxiety and often resentment. Students in preliminary interviews for this book reported that they had to yield their individual work habits and engage with the preferences of at least one other teammate. They often resisted doing so.

Framing our students' efforts in collaborative communication as a key part of their entry into their professional community helps us understand the challenges that they face. First, in collaborative communication projects, our students must write and present to a varied audience that, in addition to their professors, may also include sponsors and industry professionals, as well as knowledgeable peers. The communication is focused toward criteria for success with a nonacademic audience, and our students are often unfamiliar with these criteria. Second, students must learn not only to describe and define but also to argue and interpret in order to persuade or inform multiple audiences. Third, student writers and presenters must manage a new genre (or two or three) as they create design documents, proposals and reports, briefings and presentations, and memos and letters. Joining a professional community requires new language, a professional tone, and a new style. And finally, collaborative communication quickly reveals its dependence on team abilities. As our students negotiate their paths toward the collaboratively produced and presented deliverables, they must use a complex and perhaps not yet well-practiced set of team abilities.

Of the learning challenges we have listed, the first three (audience, rhetorical strategy, and genre) are well within the theoretical and pedagogical territory of Writing in the Disciplines and Writing Across the Curriculum. However, the strategic development of team abilities and their role in creating collaborative communication in science and engineering may seem less familiar to writing faculty and perhaps to engineering faculty as well. Thus, it is worth looking at the key team skills that are important supports for collaborative communication.

Engineering colleagues have little or no disagreement on the importance of team skills. Team abilities, often grouped with "professional" or nontechnical skills, are essential in the engineering and science professions. Moreover, "mastering [those skills] will be a major determinant of the future competitiveness of U.S. engineering graduates" (Shuman, Besterfield-Sacre, and McGourty 2005, p. 44). Loughry, Ohland, and Moore observe that a "major factor accounting for project success [is] the effectiveness of various team processes" (2007, p. 505). Other researchers have observed that team difficulties and project problems are linked, and they conclude that the quality of student written reports is correlated negatively with poor project performance and team dysfunction (Lovgren and Racer 2000, Smith and Imbrie 2007, Craig and Coleman 2004). Moreover, the importance of team skills is emphasized by their inclusion in the engineering accreditation criteria (Accreditation Board for Engineering and Technology 2000, Shuman et al. 2005).

Several researchers provide a useful starting place for defining, teaching, and assessing team skills (Ohland et al. 2005; Smith and Imbrie 2007; McGourty and De Meuse 2005). Despite this work, busy engineering and science faculty may persist in their individual definitions of teamwork. A review of the literature of engineering education research shows that *teamwork* is variously defined, and different researchers find certain facets of team skills more useful to their work than others. While it is likely that most educators are referring to the same general abilities, crafting useful learning objectives and developing pedagogy require clear definitions. Moreover, engineering faculty may persist in their accustomed methods of achieving and assessing teamwork, relying heavily on past experience or on the methods their own mentors used. And often little method may be employed at all. Accelerated engineering curricula provide little time for the study and shaping of team skills and their implementation in collaborative communication practice. Engineering faculty, pressed for time, often assume (or hope) that putting students in a group and assigning a project that culminates with a collaborative report or presentation will provide the necessary learning by what Lovgren and Racer describe as "a passive absorption of group issues" (2002, p. 158).

But researchers in collaborative pedagogy assert that "students ... do not appear ... to acquire teaming skills in the absence of structured experiences designed to develop these competencies" (Lewis, Aldridge, and Swamidass 1998, p. 150; see also Oakley et al. 2004, p. 21). Writing about collaborative communication, Lowry, Nunamaker, Curtis, and Lowry report that students collaborating on distributed work face significant "process loss" unless procedural support is supplied (2005, p. 341). Increasingly engineering faculty acknowledge that students working in teams need support (Lovgren and Racer 2002, Shuman et al. 2005).

Thus, the question that all engineering researchers and dedicated faculty would like to answer is: What support do our students need as they learn team skills? In this research, we approached that question from the student perspective, surveying and interviewing students who were part of a team deeply engaged in performing an original research experiment and who had to report on that experiment in collaborative reports and presentations.

In preliminary interviews, MIT students reported that they were no strangers to teamwork. By the time they begin college, our students have been active in cocurricular activities, academic projects, multiple organizations, and community settings. Yet these same students may struggle when they work on a demanding team project. Usually the struggle shows itself only in heightened frustration and irritability, and the teams manage to surmount these problems. But sometimes student teams experience so much conflict and dysfunction that their technical goals are compromised and learning objectives suffer. Why? Perhaps these students have not had enough instruction about team skills, although many of them have had some exposure and most would quickly say they "already know that." Perhaps they have not used those skills in such a fast-paced or competitive (grade-oriented) setting or tackled such complex projects. In addition, the process of writing and presenting collaboratively may be unfamiliar to them and feel unwieldy.

Even if students have learned about teamwork, their knowledge may not always translate into the ability to use those skills effectively and efficiently. In other words, the students may possess declarative knowledge (knowing *that*) but be uncertain about procedural team knowledge (knowing *how*), and they may lack a sense of conditional knowledge (knowing *why* and *when*) (Paris, Lipson, and Wixson 1983).

In fact, students at MIT who were interviewed and surveyed for this research repeatedly insisted that they did not need or want instruction about teamwork. They already "know that." Nevertheless, transcripts of student interviews in this research study did reveal that students had questions about how to use their existing knowledge to resolve conflict with their teammates or when and how to use conflict management strategies most effectively. Not only did they spontaneously volunteer their insights in these reflective interviews, they showed enthusiasm and confidence in their abilities to move beyond declarative or textbook knowledge. They identified project management and team skills that had been important to them and demonstrated how overall commitment to their work motivates them to practice those management and team skills well. Moreover, they revealed how challenging the collaborative communication process was for them and helped focus attention on how they would like to learn those skills more effectively.

Learning Collaborative Communication in the Department of Aeronautics and Astronautics

The research reported in this chapter was conducted in the Department of Aeronautics and Astronautics, which is organized into ten departmental laboratories and centers and engaged in approximately two hundred research projects (*MIT Bulletin* 2008). The department's pedagogy is based on the "essential functions of engineering; ... engineers should be able to Conceive-Design-Implement–Operate [CDIO] a complex engineering system in a team-based environment" (CDIO Syllabus Report 2001, p. ii). To this end, the department, with the collaboration of industry partners, has initiated the CDIO program in which the skills of contemporary engineering are codified. In addition, the program seeks to develop new pedagogical approaches to "enable and enhance the learning of these skills" (CDIO Syllabus Report 2001 p. 1).

One professor in the Department of Aeronautics and Astronautics says of the department faculty, "We are engineers who still remember how to make real machines" (E. Crawley, personal communication, July 11, 2008). As all engineers will understand, the design, building, and testing of these "real machines" (and real systems) require a great deal of collaboration. Thus, the department describes itself as preparing its "engineering graduates for ... leadership" in the CDIO effort (Department of Aeronautics and Astronautics n.d.) and emphasizes one of the four basic categories of CDIO as "Interpersonal Skills, Teamwork and Communication" (CDIO 2001, p. iii).

The emphasis on collaboration in Aero/Astro, as it is often called, is not just prescribed; it is also modeled in the department environment and fully enacted in the behavior of many faculty members. Engineering faculty often teach in teams, each contributing his or her perspective from a particular disciplinary focus. Communication instructors are part of the planning as well as of the teaching and assessing effort. The technical staff manage the lab space, machine shop, and wind tunnel as a team. The ability to do high-quality work in a team is steadily affirmed by the department's administrators. Faculty members design complex projects that mandate team approaches. Students in this department are surrounded not only by a sense of community but more precisely by multiple models of what it means to solve problems and to work together.

Working across a range of disciplinary foci, students complete their Aero/Astro major course of study with a choice of capstone projects that "serve to integrate the various disciplines" and also provide a communication-intensive (CI) experience within the student's major (*MIT Bulletin* 2008, p. 133). Both Space Systems Engineering and Experimental Projects I and II are CI team-based courses. Both require a significant amount of collaboration among students, and most of the communication deliverables are collaboratively composed and presented. In each course, the design or research challenge is ambitious and rigorous, and time constraints require that student teams work, write, and present efficiently. Table 5.1 summarizes key information about the three courses.

Table 5.1
Summary of courses studied in the collaborative communication research

Course	Semesters studied	Number of students	Communication deliverables
Space Systems Engineering	Spring 2006, Fall 2006, Spring 2007	47 (Spring 2006)	3 formal design reviews in each semester 1 collaboratively written design document in each semester
Experimental Projects I	Fall 2007	27	1 proposal, individually written in 3 drafts 1 formal presentation
Experimental Projects II	Spring 2008	24 (2 students took semester off, and 1 student dropped out)	2 formal presentations 1 collaboratively written final report

Space Systems Engineering

In the semesters in which this research was performed, the Space Systems Engineering students designed, built, and tested a Mars rover that would be able to perform scientific experiments in rigorous Martian terrain with or without direct human guidance. The three-semester course focused on achieving the stated learning objectives (LO) and specified the measurable outcomes (MO) by which those objectives would be assessed (see box 5.1).

The junior and senior students delivered three formal oral design reviews in each semester. The lengthy design reviews were collaboratively composed and presented, and student presenters were subject to rigorous questioning by engineering faculty. In addition, the students submitted a collaboratively written design document at the end of each semester, capturing the design at the end of the term. At the end of the three semesters, the design document, revised several times, was 350 pages long, not including the appendixes. The length of this document was perhaps unusual, reflecting the complexity of the mission; most design reports in this course are approximately 200 pages, including appendixes.

Forty-seven students enrolled in this class in spring 2006, although the class enrollment dropped through subsequent semesters as seniors graduated or chose alternate capstone courses. To design a complete space system, the students were divided between subsystem teams in order to complete a systems analysis; design and build propulsion and power systems; fabricate structural elements; design and build an avionics system, as well as thermal and environmental controls and support systems; and provide weight and cost estimates. A consideration of human factors was included. Thus, students not only were part of the larger team but also worked in smaller subsystem teams that ranged from two or three to eight or ten students, depending on the scope of the task. A systems team assumed many administrative tasks and guided the subteams as they integrated the various modules into one system, mirroring the authentic tasks of professional design engineers.

The teaching team was composed of engineering professors, technical staff, a communication instructor, a graduate teaching assistant, and industry sponsors and mentors. Industry sponsors attended reviews and gave feedback. Students convened in the larger class for lectures and discussion as well as for design reviews. However, students chiefly worked in subsystem teams with the support of their engineering faculty or teaching fellow or mentors. The communication instructor lectured during the first semester on technical writing and oral presentations. Student teams met with this instructor for rehearsals of their presentations and for individual or group writing conferences when the final report was due. The engineering faculty and the communication

Box 5.1
Learning objectives and measurable outcomes for Space Systems Engineering

LO 1: Summarize the mission requirements and develop a set of system and subsystem requirements that define a vehicle that meets the mission requirements.

MO 1.1: Requirements analysis in Design Document

MO 1.2: System Requirements Review

LO 2: Develop a set of Figures of Merit (FOM) that quantitatively characterize the performance of the system to meet the mission requirements.

MO 2.1: Conceptual Design section of Design Document

MO 2.2: System Requirements Review

LO 3: Develop a system architecture which provides a "best solution" to meet the mission requirements based on the FOM.

MO 3.1: Conceptual Design section of Design Document

MO 3.2: Conceptual Design Review

LO 4: Based upon the chosen system architecture, design a vehicle which: "closes" technically, i.e., satisfies the laws of nature; is build-able within the time and cost constraints; can be tested to verify that it meets the mission requirements, and is operable in the mission environment.

MO 4.1: Preliminary Design Review

MO 4.2: Conceptual and Preliminary Design sections of Design Document

MO 4.3: Report of lessons learned from previous experiences.

LO 5: Complete the detailed vehicle design with analysis and drawings to a level that the vehicle could be produced by someone else.

MO 5.1: Critical Design section of the Design Document

MO 5.2: Critical Design Review

LO 6: Fabricate or acquire subsystems and assemble the complete system to prepare for testing and evaluation.

MO 6.1: Manufacturing and Testing Plan sections of Design Document

MO 6.2: Acceptance Review

LO 7: Execute system and subsystem level test to demonstrate that the vehicle can be operated safely and achieve the mission requirements.

MO 7.1: Bench tests, system tests, flight tests

MO 7.2: Execute the mission

LO 8: Report the outcomes of the vehicle performance and resulting lessons learned.

MO 8.1: Post Mission Review

MO 8.2: Lessons learned document

Box 5.1
(continued)

LO 9: Apply project management methods to execute the project on schedule, with resource constraints, and to deliver the technical performance measured by the FOM.
MO 9.1: Management Summary section of the Design Document
MO 9.2: Project reviews: SRR, CoDR, PDR (16.83) and CDR (16.831) or AR (16.832)
MO 9.3: Risk management

LO 10: Keep records of work done and document progress made to achieve the design project objectives.
MO 10.1: Meeting notes, notebooks

LO 11: Communicate facts, findings, and ideas to peers, superiors, and subject matter experts.
MO 11.1: Informal Review
MO 11.2: Design Document
MO 11.3: Engineering Analyses

LO 12: Evaluate progress towards team and class goals.
MO 12.1: Peer reviews

Note: LO = learning objective. MO = measurable outcome. FOM = Figure of Merit; SRR = System Requirements Review; CoDR = Conceptual Design Review; PDR = Preliminary Design Review; CDR = Critical Design Review; AR = Acceptance Review.
Source: Space Systems Engineering, Program Plan (2006, p. 8).

instructor assessed the design documents and the design reviews and provided detailed feedback.

Experimental Projects I and II

Experimental Projects I and II, a two-semester capstone course in which students perform original research under the mentorship of an advisor, were the primary sites of this collaborative communication research. In the semesters in which this research was done, Experimental Projects I enrolled twenty-seven students (fall 2007). By the end of the spring 2008 semester, these students had completed projects that included experiments on reconfigurable wheels for a Mars rover, the application of radio frequency identification sensors to a storage bag to be used on the international space station, a control system for a monocopter, and the characterization of the aerodynamics of a Frisbee.

In Experimental Projects I, junior and senior students formed two-person teams. Each team developed a testable hypothesis and designed an experiment to test that hypothesis. For students, this course is often the first time they have done original research and the first time they have worked closely with an engineering mentor. Far from the "plug-and-chug" lab experiments so often done in experimental courses, Experimental Projects I and II strive to meet the following specific learning objectives:

- Develop critical thinking skills through the formulation of a precise statement of motivation, hypothesis, objectives and success criteria for an experimental study of phenomena of the natural world.
- Learn strategy and tactics for performing original research in a small team through the design, implementation, analysis and communication of an independent original experimental research project.
- Develop hands-on experience with project-specific aspects of the design, fabrication, implementation, conduct, and analysis of an original research experiment (e.g., mechanical design, machining, electronics, programming, instrumentation, interactions with suppliers).
- Learn techniques for effective project management through executing, as a team and on schedule, an experiment that successfully assesses a hypothesis.
- Learn and practice effective technical communication skills—both oral and written—for a range of professional situations including informal team meetings and formal written and oral reports. (Experimental Projects I and II Course Syllabus 2008, p. 17)

The learning objectives were designed to introduce students to independent research and the professional activities that accompany that pursuit, and although Experimental Projects I and II are classroom based and have many of the hallmarks of an academic course, most students begin to develop identities as young professionals and show this process by the ownership and dedication they display toward their research as the experimental project progresses.

The two-semester course had two distinct stages. The emphasis in Experimental Projects I was on the design of the experiment, and an iterative process was used as students thought critically about how to develop and then test their hypotheses. In Experimental Projects II, the team built the experimental setup, performed the experiment, collected and analyzed the data, and assessed the hypothesis.

Experimental Projects I concluded with an individually written design proposal from each student. Despite the emphasis on collaboration in this course, the engineering professors and the communication instructor deliberately preserved the individual writing of the proposal, feeling that students are better able to collaborate on a communication deliverable once they have some experience in creating a whole technical document. Students wrote this proposal in three carefully structured stages. First, the hypothesis, objective, and success criteria were collaboratively written by the team. Second, each individual student added an introduction, a review of relevant literature or theory, and an explanation of the proposed technical approach. Based on the critique from the engineering faculty and the communication instructor, each student revised the early version of the document. Third, each student added a section explaining the experimental design; a discussion of data collection, error analysis, and data analysis; a project planning section; a list of sources cited; appendixes; and the executive summary. The objective of the proposal was to persuade the faculty and technical staff that the experimental design was robust enough to collect meaningful data that could then be used to assess the hypothesis. In addition, based on this material, the student team gave a collaborative oral presentation of their work. This presentation was subject to rigorous questioning from the engineering faculty, communication instructor, and technical staff.

For students, the last phase of the proposal and their proposal presentation was an introduction to an unfamiliar professional genre. As was the case for the students in chapter 3, many students were unaccustomed to making a persuasive case for proposed work, feeling perhaps that their excellent or innovative idea was evidence enough to persuade a reader. Their early drafts of the individual proposals and their early drafts of the collaborative presentation sometimes sounded more like an enigmatic report than a persuasive document. In addition to learning to persuade, students began to assume a more professional voice in their writing and in their speaking as they worked through the proposal process.

In Experimental Projects II, there was no individual writing. Student teams gave two collaborative presentations during the semester. They began the semester with a short oral progress report in which they were expected to present the current status of the project, raise any problematic issues, and offer a plan to resolve those issues. At their second presentation at the end of the term, teams reported on the data they had collected. Teams were expected to argue persuasively for their interpretation of the data and for their conclusions as they assessed their hypothesis. Each team concluded its work by submitting a thorough and substantive design report. Again, the Experimental Project II final report was an opportunity for students to practice a professional genre, a professional voice, and to do so collaboratively. Often students have used their final presentations as the basis for conference presentations.

The teaching team was composed of engineering professors, technical staff, a communication instructor, and a graduate teaching assistant. The communication instructor lectured on technical writing, oral presentations, and graphics and gave workshops on how to compose presentations, the design proposals, and the final report. Students also met with the communication lecturer for writing conferences before documents were submitted and for rehearsal before presentations; and they met with the teaching team and their advisor for presentations and team meetings.

Key Findings: What Students and Engineering Faculty Say about Learning to Write and to Speak Collaboratively

Our focus groups, surveys, and interviews support five key findings about how MIT engineering students learn collaborative communication, which team skills help them do that work, and how they best learn (or want to learn) those team skills:

- Project management and team skills are as important to the collaborative communication process as they are to the processes of design and research.
- Commitment to an overall project goal motivates students to develop and strengthen team skills.
- Students who are new to or inexperienced in the collaborative communication process benefit by explicit structuring of that process.
- Students and engineering faculty appreciate the importance of team skills, but there is little consensus on a single "best" method of how to achieve those skills.
- Students find specific verbal and written feedback helpful as they learn to communicate collaboratively.

Data from surveys of students and faculty and excerpts of student interviews with five students illustrate these five findings. In this chapter, we have chosen to look at findings across individual cases rather than at individual students, a decision based in part on a methodological limitation. Not all members of the teams studied participated in this study. Therefore, in order to maintain the confidentiality of the students who did not participate in this study, we could not focus on the products of a team. Instead, in this chapter, we use the experiences of students—Barbara, Karen, John, Peter, and Kaya—from various teams. Barbara, Karen, John, and Peter were interviewed over two semesters, and Kaya was interviewed once at the end of the Space Systems Engineering sequence. Barbara, Karen, and Kaya were seniors, and Peter and John were juniors. Both Kaya and Peter had experience with team building and leadership in a military program, while Barbara, Karen, and John had chiefly the team experiences of their undergraduate courses. John was also active in his fraternity's projects. For Barbara, Peter, and John, the following five findings were consistent themes in the sequence of four interviews. Karen also echoed these findings, although, for reasons discussed later, not at length. Kaya was especially focused on project management issues because of the circumstances of her collaborative work.

Project management and interpersonal skills are as important to the collaborative communication process as they are to the processes of design or research.
In the 2007 focus group surveys, students ranked project management skills as "somewhat important" but not "very important." Yet in the discussion that followed the surveys, a few students volunteered their difficulties with time estimation and planning. One said, "It was hard for me to estimate the time that a task would take. This made planning and scheduling hard. I learned to build time into my planning for schedule slippage."

Students also had difficulty in dividing work and reflected on what can happen when tasks are delegated to others:

"We do not know how to distribute work."

"If a person falls behind and does not learn [a skill set] at the beginning of a term, then it is too late for him to catch up. And then that person has to assume a less critical role. So the work distribution is uneven."

Some students had only a vague sense of what project management was and how it was put in place. "It just happens," one student commented. "Leaders emerge organically," volunteered another. A third student commented (somewhat optimistically), "Get to know one another and project management happens."

Similarly, a year later, students in the Experimental Projects I and II capstone and their engineering faculty and mentors were surveyed and ranked project management skills as only "somewhat important" but not "very important." However, in contrast to the survey data, students in interviews emphatically identified specific project management skills that were very important to their teamwork: (1) division of work and then the re-combination of that work and (2) estimating time, planning, and scheduling the plan. In addition, they found that decision making, conflict resolution, and interpersonal skills were vital to their success.

Division of Labor and Communication Tasks Engineering students are understandably concerned about dividing work. Initially enthusiastic because of the efficiencies that shared work can provide, students often retreat to their already established individual work patterns. Some are simply unused to sharing work. They prefer individual effort because they feel more in control. And some claim that individual work is more efficient because they do not use up time talking about it. However, tackling large, complex projects forces students to divide work in order to succeed. The difficulties in work division often come from feeling out of control of the process and not knowing whether their teammates have the necessary skills or accountability to accomplish the task. With high-achieving and competitive students, egos sometimes can become involved.

When the work is divided and accomplished, students again can be inexperienced at (or mistrustful about) recombining the work into the final product. Not only does recombining shared work require planning and scheduling, this stage also reveals how well the teammates communicate with one another and how reliable and thorough each of them has been. Excerpts from John's interviews illustrated some of this struggle.

John began his first interview by talking about the way he and his teammate planned to divide the work: "Terry has more extensive knowledge of aerodynamics than I do ... so he's helpful in explaining some of the things that are going on.... And my role ... he's an exchange student ... so I help him grapple with some of the terminology and with some of the technical jargon. We'll start out individually and then come together and share our work and try to come to an agreement."

With much discussion, John and Terry managed to divide the work, but they had difficulty recombining that work effectively. At the end of the semester, their method resulted in an unsatisfactory project presentation. In his second interview, John explained, "I had to be out of town the weekend before so we basically split the work of the oral presentation in half.... We did the two parts independently. We looked at each other's slides the day before but we did not revise ... there were things in both parts of our presentation that could have been corrected had we worked a little bit more together on the presentation."

John continued to talk about his conflicted experience with shared work and how difficult it was for them to decide who would assume which task: "I think delegation is very important.... But then sometimes it is just hard to compromise on how things should be done. One decision can take a long time to make."

By the beginning of the second semester, when the project work had intensified, John and Terry had divided the work differently. John, somewhat hesitantly, reported, "This [delegation of duties] has been successful. I think the work is still getting done."

In the fourth interview and at the end of the project, John, talking about conflicts in the project, showed increased insight into how the team's lack of agreement about work division fueled those conflicts:

> I wanted to be able to separate work. You know, "Terry, you take care of construction, I'll take care of data collection." I saw we were not getting a lot done when we were always working

together. . . . But Terry insisted on and even through the first semester, he was insistent upon us doing work always together and that's one of the points where we clashed. Terry wanted to be involved in every step of the process. I thought we could work separately and then summarize it to one another.

John went on to describe how his plan affected the collaborative writing of their final report:

We delegated the sections and . . . went off and independently wrote sections for our final paper. When we came together . . . , that's where the problems started. There was really a sense of ownership in our writing . . . we were reluctant to have the other person edit the writing . . . not so much grammar but structure. There was a sense of . . . "this is mine." We did not get past that until later than I would have liked to. And that's . . . probably the biggest problem we had in the course . . . letting go of the individual ownership and learning to think in the collective sense.

In his interviews, John demonstrated his ambivalence toward the collaborative project as well as the collaborative communication. He wanted to divide work for greater efficiency, but the interviews revealed his ambivalence and hesitancy about the actual accomplishment of the work. By the end of the study, he talked openly about the difficulties with work division that had consistently arisen during the course. John and Terry had discussed this pattern with the teaching team at several meetings, and by the end of Experimental Projects II, John had gained some insight into his conflict with Terry. But the effort devoted to struggling over a basic management skill sapped the team's energy.

Conversely, Barbara reported that she and her partner, Amy, had few problems with work style and division of work. Friendly acquaintances for several semesters, Barbara and Amy worked well independently and were skillful about combining their efforts collaboratively. In her second interview, Barbara said, "Maybe we have individual work to accomplish and [then] we'll come back at a later point and . . . 'hey, this is what I found.' And because of that, when we got together to make our presentation, we both knew what was going on. So by consistently working together and having meetings where we discussed what was going on, we were both at the same level when we put the presentation together."

By the beginning of the second semester, Barbara reported that she and Amy were still using a work style that allowed them to efficiently divide work but also to work together. Their project management style seemed egalitarian and relatively free of issues of ownership: "There are times when I feel like the leader and there are times when I feel that I am listening to Amy's suggestions and trusting her, what she is saying. And other times, we're both just working together on the same level, on the same task."

Barbara and Amy's ability to not only divide work evenly but also to recombine work efficiently through their consistent communication process clearly contributed to the efficiency and effectiveness of their team. Both students enjoyed their work together and made steady and impressive progress on their project.

Estimating Time, Planning, and Scheduling the Plan Another key project management skill for teams is the ability to plan and schedule. Planning begins with work division but includes learning to estimate time needed for tasks. This estimation then forms the basis for a feasible schedule. Engineering students often do not have the experience necessary to estimate time well. Also, they are accustomed to working individually, and so they sometimes neglect to account for the difficulties of synchronizing work with another or several others. In her fourth and last interview, Barbara reflected on her team's ability to plan and to schedule:

> I think being able to look ahead and plan things well was very important.... If you aren't planning things well, then you're not meeting deadlines, then it is easier for your team relationship to break down. It was not like "we have to do this for Experimental Projects II" so we pull an all-nighter and then did not look at it again for five days.
>
> We said, "ok, we have a five hour block of time here so we can work on this," and then we have a chunk of time tomorrow or the next day so we sit down and work on the next thing.
>
> ... The all-nighter strategy gets ingrained but that strategy does not work for a project like this ... you are not putting yourself in a place to succeed.

Kaya, the managing editor of the design document in the spring 2008 Space Systems Engineering semester, compiled drafts from fifty-two writers and then, with the help of a small team of writers, edited and revised a 350-page document at the end of the term. When asked about the project management skills needed for collaborative communication, Kaya immediately identified the ability to plan: "One of the biggest things is long-term planning and being able to think past the next deadline. For example, the design document was not due for a month and a half, but I looked at the calendar and realized 'oh gosh we need to have people in place....' Specifically we had someone who—every week—would stand up and say 'these are our deadlines for the next three weeks.'"

But project management skills were not the only important skills for the students interviewed. Most of the students identified team skills (in this research, decision making, conflict resolution, interpersonal awareness, and communication) that were vital to their success.

Decision Making, Conflict Resolution, and Interpersonal Skills In Experimental Projects I and II, each team had to make many decisions small (when to meet and where) and large (how to design their experimental apparatus). Their preparation of collaborative reports and presentations offered multiple opportunities for decisions too. Perhaps because the teams were small and students were anxious about encountering conflict in such small teams, most of the students interviewed saw decision making in terms of achieving consensus with their partners. Generally their consensus-based decision making took place after extended negotiation. For example, in his second interview, Peter talked about compromise: "We're pretty different as far as getting things done. So it is just try to adapt to one another . . . and to the presentation you're trying to prepare. . . . It is a small thing but I wanted to do a list one way and he looks over my shoulder and says 'why are you doing it that way? You could do it this way.' I said 'tell me why.' And he gave me his explanation and I said 'OK, that sounds good' so we switched over to his way . . . we compromised."

In his second interview, John also connected decision making with compromise but indicated that these compromises were hard to obtain: "[A choice about] something small can generate a lot of debate so the ability to compromise is important when writing these technical documents." John continued to describe the negotiations that preceded compromise: "With MIT people, we're smart but we are also very stubborn and when we think we have the right answer, we are very, very unwilling to back down. I think being able to realize you might be wrong is an important part of working in a team."

Kaya, who was working with larger teams of writers, had less opportunity to reach consensus. She decided (with faculty support) to impose her authority on the writers in order to meet the deadline. Her peers expected her to make different choices about them and the sections they were submitting. Kaya said, "[They] think 'oh she's my best friend so she's obviously going to include my section no matter how late I am.'" Because of her military background, Kaya saw her decision-making and policy-setting role as being less connected to compromise, negotiation, and consensus than to a chain of command. She compiled the draft and sent it to the faculty with sections left blank where the writer had not complied with the deadline. She reported that her classmates were harshly critical of her choices, and in at least one case, she and an upset peer were in conflict with one another for several weeks. This was unpleasant for Kaya, but trained in military leadership, she realized that being in authority is often difficult. Despite her discomfort with the negative feedback from peers, Kaya's prior experience in leadership and command positions influenced her ability to take a stand professionally; it was part of her job, she said.

When decision making was difficult and team members could not reach a compromise or when conflict arose, then conflict resolution was the skill that helped a team succeed. John and Terry, because of their steady disagreements about work division, had a lot of experience with conflict resolution. As early as the first interview, John volunteered that he and Terry were at an impasse: "There are a few rough spots ... we did have some trouble coming to agreement on some of the parameters.... And we still haven't reached agreement. I think we're going to have to talk to [their advisor] about it because it does not seem like either of us is budging." But by the third interview, John reported that he and Terry had resolved some of the conflict and were better focused on the project: "That friction between us has gone away. We're both focused on getting data. And a lot of that was just figuring out how each of us worked.... There are still times where we disagree on things but ... it is much easier to work through as opposed to last semester when it was kind of like feeling each other out."

By the end of the course, John reported, "Conflict resolution skills are what's evolved most over the course for me." He said,

> I definitely think my skills have improved over the past eight or nine months. I think I really did not understand how to deal with people.... I did not really understand how to resolve conflict in a team situation because it hadn't really come up in the teams that I had been on before and if it did come up, people just kind of brushed it off. It was not confronted directly. It was confronted much more directly in Experimental Projects I and II. I have a better idea of how to handle conflicts whether it is laying out reasoning to the other person and walking through the thinking that got you to that conclusion or, if you have to, going to someone higher up to have them mediate the problem.

Not surprisingly, interpersonal awareness and interpersonal communication turned out to be important to the student teams too. A survey question asked about the importance of interpersonal awareness and the awareness of the psychodynamics and learning styles of others. Both students and the faculty members surveyed scored this item as "not very important." Yet once again, students gave a different response in the interviews, spontaneously reflecting on their better understanding of their partners and themselves.

In contrast to their ranking of interpersonal awareness, both students and faculty rated interpersonal communication (active listening and clear communication) as "very important." In interviews, Barbara, John, and Peter agreed, directly connecting their increased interpersonal awareness with improved interpersonal communication within their teams. Moreover, it was not only that each student was aware of his or her partner; students also felt they understood themselves more deeply. In her fourth interview, Barbara described her teammate's strengths and weaknesses as well as what she had come to learn about Amy's particular personality style: "I've come to better understand where her technical strengths and weaknesses are and also just a bit more about her own personality and mentality ... all of which I think were important to be aware of in order to keep our team relationship healthy." Barbara went on to describe her insight into herself as well, and she noted how this insight helped her avoid conflict and how the team dynamic had improved:

> I think I've learned to be a better listener. Sometimes when you're working with a teammate ... I think it is easy to listen ... but not really take it seriously because in the back of your mind, you think, "oh well, I know more about this than she does." ... What I learned was how to actually set that bias or stereotype aside. I found out that when I did that it made our team dynamics better.... She felt like she could express herself and there were times when she had a very valid point and by being able to set my own bias aside and really listen, ... it strengthened the final product.

In his second interview, Peter described his increased understanding of Mark and the positive results for their work together: "As time goes on, we are starting to understand one another a little more ... so there's more understanding ... and more patience.... For example, he's maybe not as fast a writer as I am ... but we deal with those differences together. We find ways to work on it."

Summing up in his third interview, John, while reflecting on some of his difficulties with his teammate, said, "You have to find out how people 'work' before you can work very efficiently."

In summary, students in this research spoke openly and reflectively about what they had learned from their team-based project. As might be expected, none of the students came to the Experimental Projects I and II experience untouched by previous experiences or by their own maturational processes. The interview transcripts show clearly that each student had specific strengths or was challenged by particular gaps in his or her abilities. Although the students were having parallel team experiences, each student was actively and privately engaged on his or her own individual developmental continuum. Barbara described becoming a better listener. John reflected on his improved ability to resolve conflict. In his last interview, Peter contributed that "the skill that I used more often any other [was] being open with [Mark], being patient, being willing to listen to ideas." Kaya practiced the command skills that would carry forth into her military career, and she worked on feeling more at ease with that authority.

However, not all students experienced such insights. A fourth student interviewed, Karen, did not identify specific insights into herself or her partner nor did she identify project management skills that she had learned or strengthened. She and a friend, Charles, had formed a team, and she reported no difficulty with decision making or conflict resolution. Karen and Charles had been close friends for three years, and both belonged to a strong network of mutual friends. In her third interview, Karen said, "Another project might have been more interesting to me, but I know Charles, so I went with the project that he wanted to do." She described the other courses in which they had worked together and ended by talking about their similar interpersonal styles:

"As far as decision making goes, we see eye to eye. We might initially have differing opinions, but we end up agreeing. There are very few people I could work with as well as I work with Charles." Thus, in situations where other undergraduates might struggle for control over the project or be frustrated about a teammate's work style or argue over a decision, Karen and Charles appeared to work smoothly. Their project was successful, and they obviously enjoyed their work together.

The experience of Karen and Charles is often reported when student teams choose their teammates based on a strong friendship. In a sense, the existing friendship creates communication patterns into which the team process meshes, and the technical work can begin more swiftly. The relationship has also taught each teammate about the working style and the personality of the other. They either resolve conflicts quickly or avoid them altogether because the teammates know one another so well and because the relationship itself has a high priority. Often for student teams in which the friendship is the basis for their team formation, the team process seems to flow easily and effectively. Students in teams like these rarely see the need for discussion or teaching about team process. Teammates like Karen and Charles usually report satisfaction in their project work. But although they may not experience the sometimes uncomfortable stages of a developing team, they also may not develop the rich awareness and enhanced collaborative skills that Barbara, John, and Peter attained.

Although student survey responses tended to neglect the importance of project management and interpersonal skills in collaborative work, student interviews gave a vivid and detailed picture of the essential nature of those skills. Moreover, interviews documented the ways in which these skills were integrated in these collaborative projects and the ways in which each student developed individually.

Commitment to the overall project goal is a strong team motivator.

What motivates a student team to work together successfully, overcoming challenges, mastering team skills, and producing high-quality work? In Experimental Projects I and II, it did not seem to be the grade alone, although these students are high achievers. It did not seem to be the exhortations of the advisor, although students valued their advisor's opinions and mentoring highly. Interpersonal team relationships were influential, of course. But a key element that helped team members develop and perform seemed to be a commitment to the project.

Barbara, Peter, and John could identify the point at which the project became "their" project, not simply a school-based process of fulfilling a course requirement. In fact, a strong commitment to the overall project goal is a characteristic of high-functioning teams, as many researchers observe. McGourty and De Meuse describe it as a "unified commitment—members put the team goals ahead of individual needs" (2001, p. 8). Katzenbach and Smith (1993) echo this and add that the team goal must be a challenging one if the team is to coalesce around it. Greitzer describes "an emphasis on the overall project goal" as essential to the team's success (2007, p. 15). Widnall, quoting Ed Schein, an organizational psychologist, claims that true collaboration happens only when there is a "common task at the boundary." She describes this common task as something that is important to all partners and cannot be accomplished alone. Moreover, "You have to believe that a decision made in consensus with a team is the better decision," she claims (S. Widnall, personal conversation, December 18, 2008). Most of us who work in collaborative efforts understand these concepts well: the sense of ownership of the project as a whole and the dedication to the project's success, the interdependence among colleagues, the mutuality that promotes conflict resolution. We are motivated by our common goal. Without that motivator, our collaborative energies may flag.

Students in the Experiment Projects I and II interview group discovered this phenomenon for themselves. Peter said in his third interview: "For us, one of the most important things was having a common goal . . . a vision for the final project." Barbara in her third interview described her mutuality with Amy: "There was [increased] harmony between my partner and me as far as really understanding, thoroughly understanding our project on a mutual level and having a mutual goal of where we want to take this." In her last interview, she reflected on the commitment she and Amy shared: "What helped was not having a focus on ourselves but on the larger project . . . as it turns out, neither one of us is very arrogant." John, in his last interview, reported on the mutual commitment that he and Terry had begun to experience: "At the end of the project, there was much more willingness to compromise in order to reach our goals."

Dedication to the overall goal can be affected by other factors, and it does not always develop. Sometimes students hastily choose a project that ultimately does not interest them that much or they are forced to join a project that is outside their area of interest. Perhaps they lack the skills sets necessary to succeed. Several of the students who were surveyed in Space Systems Engineering speculated about why that commitment to an overall project goal may not develop, thus affecting the formation of a strong team. Sometimes the structure of the course is not easily understood and students do not grasp the main objective. One student commented, "Working in a small subsystem team, we lost sight of the big design picture." For another student, location was problematic. He said, "We were too far away. Moving [to a closer lab] helped us feel connected to the project." Sometimes a team is too large for a student to find a niche at first. A student said, "There were so many people . . . in the first semester, I focused on myself; in the second semester, I focused on my team; by the third semester, I became loyal to the whole project." Or perhaps the scope of an ambitious project is too large for the student to encompass. A student said, "[The project had] too big a scope at first. Once the project became more concrete, I felt more loyalty to the larger project."

A few students in Space Systems Engineering had difficulty achieving dedication to the larger project goal, although many students did not. Conversely, however, the students interviewed in Experimental Projects I and II described a more positive experience in this respect. The Experimental Projects I and II students were working on projects of their own choosing and were closely involved in all aspects of experiment design and implementation. Working in small teams, they received a great deal of feedback from the teaching team and their project advisor. Thus, it seems possible that learning to work collaboratively—whether on communication or design or experimental research—might be positively or negatively affected by elements in course design and assignment design. When these elements are well crafted, students' chances of developing a sense of ownership of their project improve greatly, and their perception of the value of their role within that project improves too.

Students who are new to or inexperienced in the collaborative communication process benefit by explicit structuring or scaffolding of that process

Moving into a collaborative communication process was not easy for the students interviewed. Barbara, Peter, Karen, and John all described strong individual writing processes that were roughly similar: some kind of structured invention or research, a drafting process, and an editing and revision process. When they spoke about individual humanities-based oral presentations, they reported a less rigorous composition process, little editing, and very little practice or rehearsal. Yet all agreed that presentations and writing in their engineering courses required more rigor and put more pressure on the writing or composition process.

Peter and Karen had had some experience in presenting collaboratively in a prior engineering course, so they had a few beginning ideas about how a group of students could put together and rehearse a technical presentation. Moreover, Peter had had experience with a military organization, so he was skilled at preparing briefings, albeit as an individual. However, when asked in interviews about how collaborative communication work was divided or how the process worked most efficiently, neither could offer a clear explanation. John and Barbara were less experienced, having chiefly the individual writing and presenting experiences typical of most undergraduates. Kaya had some experience in a summer internship. Thus, although their levels of experience varied, the students were still relative novices at the collaborative work that lay ahead.

Tightly Structured Collaboration in Experimental Projects I and II Experimental Projects I and II are courses that lead the students through the sequence of designing and completing a research project. They are also guided through the collaborative communication deliverables by a structured process that is gradually reduced. Thus students in Experimental Projects begin by writing individually, but they end by presenting and writing collaboratively.

In Experimental Projects I, each student had completed the early stages of his or her proposal. For the collaborative review, each team had to combine their ideas and present its work up to this point. Moreover, the presentation time slot was only thirteen minutes long. Thus, the teammates had to focus on the relevant material for their audience of professors and peers; they had to achieve a mutual professional voice and style. Working with PowerPoint presentations, student teams had to learn not only how to create complex graphics but also how to present them.

To structure this undertaking, the communication instructor first lectured on oral presentations of technical material and made suggestions about the collaborative composing process. Special attention was given to the preparation and briefing of the team's graphics. Each team then drafted a presentation. Rehearsal (or dry run) times were scheduled for each team, but teams were also expected to practice independently. Rehearsals took place with the graduate teaching assistant and one or more other teams present to simulate a small audience and also to allow students to learn from one another. The communication instructor used the faculty's grading rubric to guide peer feedback and help presenters focus on the key expectations of the faculty.

As each team practiced, their early efforts at explaining their complex projects were, not surprisingly, uneven and sometimes clumsy. Sometimes the draft presentations lacked sufficient or precise material, making it difficult for the instructor to comment usefully. For example, John and Terry had written parts of their presentation separately, so some parts of the presentation were underdeveloped, and neither was entirely sure what his teammate had included. At their rehearsal, they were given feedback and suggestions. However, their final talk was still scanty and disorganized because they had not allowed enough time to develop their ideas, combine the material well, or practice their presentation. After they presented, John and his teammate watched several of their peers' presentations and received written feedback from the engineering faculty as well as from the communication instructor. Based on this input and their own assessment, they understood the cause of their adequate but not very successful presentation. Barbara, Karen, and Peter (and their teammates) adhered more closely to the suggestions given, and they created and practiced their presentations together. The engineering faculty and the communication instructor assessed these presentations as much more successful, and the teams' grades reflected that success.

However, it was not until five weeks later that some of the results of this explicitly structured instruction were observed in independent work done by Barbara and Amy. During the January break, Barbara and Amy were invited to present their work to a highly respected industry partner that also collaborated with their project advisor. The stakes were high: the students were representing not only their work but also, in some respects, their advisor's work. In their presentation, Barbara and her partner not only integrated the earlier instruction into their team process but also had adapted it to fit their own scheduling demands. Each student had obligations that took her away from campus during the January break, but their robust process of outlining and developing content areas and their careful planning and scheduling and frequent communication kept them both focused. Barbara and Amy not only had reused some of their presentation from the first collaborative talk, but they had—based on their audience analysis and a much expanded time slot—used preliminary data to create more effective graphics. Moreover, they had storyboarded these graphics so that each graphic worked coherently with the next. Both were able to brief each graphic thoroughly, thus contributing to the clarity as well as the substance of their presentation.

Barbara described the deliberate process by which she and Amy planned and created their presentation, and she could identify the way in which collaboration had strengthened their work. In her fourth interview, she recalled: "We actually sat down and started talking about 'ok what is it that we want to present' because there are so many different ways we could present the data that we were seeing. We both had very different but very good ideas about how to present that data.... By combining our ideas, we came out with a product that was a lot better than if we had done things the way I thought or just the way Amy thought." Although faculty reviewed the presentation once, the students edited and revised it several times on their own. Finally, Barbara and Amy practiced independently, each critiquing the other's performance: "We listened to one another and I'd say, 'You did this when you were presenting that part and I think it would be better if you did it this way' ... and she would do the same thing for me." The team gave their presentation to the industry sponsors, and the feedback was positive. In a little more than a semester, Barbara and Army had gained the perspective and strength to present at a professional level, according to their project advisor.

Peter and Mark had a different collaborative presentation experience that revealed, in its own way, how well the explicit structuring of the collaborative process supported their presentation. Both students collaborated on a presentation and rehearsed with the instructor as well as with some peers. Their goal was to be able to demonstrate their experimental apparatus during the presentation, and they had worked toward this objective steadily. Peter and Mark had practiced their presentation, each critiquing the other and each studying the presentation and talking his way through various slides. At the last moment, because of a scheduling misunderstanding, they found themselves rushing into the presentation room with only seconds to spare. Mark was temporarily flustered, but Peter, because of their careful combined preparation, was able to take over the presentation smoothly, presenting not only his part but Mark's part as well. When it came time for the demonstration, Mark easily assumed his role, successfully demonstrated the apparatus, and answered technical questions. Their advisor and the faculty evaluated their presentation as highly professional. Peter and Mark felt that they had succeeded as a team.

Reducing the Structure in Experimental Projects II After the closely structured work in Experimental Projects I, students proceeded to Experimental Projects II, where there was deliberately less structure. While Experimental Projects I included many lectures and multiple communication opportunities, Experimental Projects II was designed to be as much like a professional experience as can be achieved in an academic setting. Student teams were given the freedom to schedule and implement their research independently, although there were set points at which they met with the faculty, participated in a class, or took part in a workshop. But unlike Experimental Projects I, where the work was structured closely, students in Experimental Projects II were expected to take the skills taught and learned in Experimental Projects I and transfer that knowledge to their semester of experimental work.

Students in Experimental Projects II gave two collaborative oral presentations: a progress report early in the semester and a final report in the last week of the semester. While these were generally successful, some students reported stress, confusion, and sometimes conflict. Because these students were spending most of their time in lab and relatively little time in class, they did not receive a repetition of the instruction given in Experimental Projects I. And although the communication instructor offered rehearsals for both the progress report and the final report, only a few teams set aside time to rehearse. Several teams did send drafts of their presentations to the instructor, as they were invited to. In general, though, the students were preoccupied with getting the experiments set up or had just finished taking data and were absorbed in analyzing those data. Thus, they neglected to schedule for themselves the practice and revision cycle that had been structured for them in the previous semester. As a result, the progress reports were often inconsistent. While some student teams understood the rhetorical purpose and the context of the progress report, other teams delivered unorganized talks that showed little understanding of their audience's purpose. The final oral presentations were generally successful, although the faculty observed rough spots in some presentations where they thought students could have refined the content or the presentation organization or delivery.

Therefore, it seems that when students are new to the collaborative communication process, structuring the process helps them achieve a professional standard. When the structuring is reduced, the students who have not internalized those habits neglect the very practices that have produced success. Of course, all structured learning must eventually be reduced to allow student growth, but the useful question is how gradually it must be reduced in order to preserve earlier gains.

For the Experimental Projects II students, the largest collaborative communication task came at the end of the second semester: writing the final report. In this report, students were encouraged to use sections of their earlier proposals in addition to the collaborative graphics developed for presentations. However, writing a forty-page document collaboratively was a challenge. Students had to decide what previous materials to use and then had to blend the work of two students together into a report with one voice. Some sections had been developed and were relatively unchanged, but the more critical sections—data analysis and the graphic representation of that analysis—were being composed. Moreover, students were approaching the sections of the document with more challenging rhetorical demands: evaluation, interpretation, and argument

In addition to posting model reports, the communication instructor described several forms of collaborative writing: the horizontal division model, the sequential model, the stratification model and the various modifications possible among models (Ede and Lunsford 1983; Ede and Lunsford 2001), Stratton 1989, Michaelson 1990, Schulz and Ludlow 1996). Students decided on the collaborative writing process that they would use. Then student teams conferenced with the instructor over their final drafts.

In most cases, the collaborative final reports were evaluated as "well done" by faculty who, using a rubric, evaluated the experimental design, the methodology, and the interpretation and discussion of results as well as writing skills. Yet despite their successes, students reported that the process of collaborative writing was stressful and often inefficient, and sometimes full of conflict. In his last interview, Peter said about his collaborative writing process with Mark, "You see your partner writing something you do not think will lead to a successful . . . paper and you're thinking 'do not do that! You're going to bring my grade down!' That pressure can get to you." Barbara said she thought the collaborative composition process had been stressful at times. She said in her fourth interview, "You're depending on the other person and it is really important ... and there were times when I was frustrated with her ... we were not working enough."

Structuring of Collaborative Communication in Space Systems Engineering In the larger capstone course, Space Systems Engineering, the explicit structuring of collaborative communication had different degrees of success. In this course, with larger teams of students and a common project with a large scope, the structuring of the collaborative presentation process was more effective than that of the document writing. Moreover, the collaborative presentations were generally more successful when assessed against the criteria developed by the faculty: clarity, coherence, consistency, technical substance, graphics, and presentation style. The presentations were design reviews that pinpointed design thinking at specific points. For example, in the first semester, small teams of four to six students briefed the faculty and their peers on the system requirements for the design mission. The second review was about the preliminary design. The third was the critical design review: it was at this time that the design of the vehicle to be built was finalized.

The engineering faculty gave the students a specific outline of their expectations of what each of these reviews (modeled on industry standards) would include. Then students drafted the presentation. The communication instructor, along with a graduate teaching fellow, watched a rehearsal of the review and provided comments and feedback; active discussion on the content, the delivery, and the composition of the presentation followed. Thus, student teams had had a lot of detailed feedback and practice before they presented to the engineering faculty. Not surprisingly, the presentations tended to be close to a professional level.

The effort that went into structuring the process of making the first presentation was useful in another way. The first collaborative presentation acted as a model for the subsequent presenters. Since all the other students watched the first presenting team, they had the benefit of seeing students tackle the same communication task that they themselves would have to address. The audience of students learned from the strengths as well as the weaknesses of the presenting team. And when it was the presenting team's turn to watch, they reinforced what they had learned.

The structuring of the design document process was less successful. Although the instructor met with Kaya and several of the writers and also provided editorial comments on the first draft, the instruction that was clearly effective in the smaller teams did not seem as effective in the larger teams working on a much larger document. Moreover, student interviews showed clearly that project management of collaborative presentations and reports was not intuitive. For example, Kaya, who led the editing process, felt that the faculty left the structuring of the process for the collaborative design document until too late. Students did not understand the process, she claimed: "If students could see the whole planning process and on day one, the professor said 'you are going to turn in a 350 pp document on the last day and this is how it is going to get done' and then take the students through the process . . ."

If the students had perceived the complexity of the collaborative task or if they had had professional experience, they might have been motivated to organize earlier, but since they were unfamiliar with such a large and complex collaborative task, they did not perceive the planning and scheduling issues. Moreover, since the design document was so large, students did not seem to have a sense of ownership about the document. They wrote their section (when assigned), turned it in, and hoped that someone else could compile and edit the document well. The sense of mutuality documented in Experimental Projects I and II was rarely articulated or observed in the collaborative writing of the Space Systems Engineering document.

In Space System Engineering, it seemed that the structuring that was provided came too late in the term, and it was not comprehensive enough to motivate a large group of students. The students' allegiance was to the design of the vehicle or their team and less so to the document itself. Thus, when Kaya and her editing team tackled the process of compiling the document, the necessary collaborative process was not solidly in place. The stronger interpersonal relationships were within subsystem teams rather than throughout the overall class. Although the editors turned out a useful document, the process could have been much more efficient and effective.

In summary, the research revealed that students benefit from explicit structuring of collaborative communication. They can take the structured management and review practices provided by faculty and use those tools to create strong presentations and documents. Whether those skills become fully integrated into a student's individual skill set is not guaranteed. And it is not clear that students always intuitively transfer the process they learned in one course to the deliverables in another. Moreover, a busy student struggling to finish a demanding project is not likely to ask for increased structuring even though he or she might benefit from it. Finally, the scope of the project, the sizes of the teams, and the pace of the technical work can directly affect team functioning. Faculty members may not be able to control team dynamics once set in motion, but before the project begins, faculty can reflect on the appropriate scope of a project, the sizes and composition of teams, and the pace of the technical work, as well as the assessment methods to be used. Reflection on these elements can help in determining how and when to provide collaborative communication instruction and how to structure that process.

While students are interested in improving team skills in the context of a specific project, and faculty members fully appreciate the importance of team skills, there is little consensus on a single "best" method to achieve this result.

In this research, one survey item or question received a consistent answer: when asked in focus groups and in surveys how students best learn team skills, students consistently answered that they do not learn how to be better team members from lectures, reading, or handouts. Students' interviews supported this claim: lectures on what they describe as generic teamwork material are not useful, they said. In a survey, engineering faculty agreed with students, although perhaps not as strongly. An engineering professor commented that "[team skills] seem to be something best learned via actual experience although some formal instruction would be helpful. But it is a constrained system i.e. we can only teach so much. Although team skills are extremely important, I would not want to see time taken away from technical skills."

Yet once again, student responses in focus groups and interviews gave a richer description of what students want and need as they learn to be stronger team members. In the 2007 focus group discussions, students consistently replied that they learned about team skills from their mentors and also from watching more advanced students. In those focus groups, students (notably at the end of a three-semester project) said they "learned teamwork by doing teamwork." A graduate teaching assistant agreed that students learn team skills by direct experience and lots of it. "It is total immersion," he reports. Close enough to his undergraduate career to remember those early stages, he went on to say, "Lectures do not help. You have to learn to build relationships before you work with those people" (B. Holschuh, personal communication, November 21, 2008).

Although students did not think lectures were useful, the survey responses indicated their interest in faculty support and dialogue around team issues. One student said, "Don't give [us] lectures but just be present as a resource." Another student urged faculty to "take some kind of mediation role especially if conflict gets bad and cannot be resolved by the team." Several students echoed their request for concrete help: "Give us positive suggestions for how to resolve conflict"; "show us how to do better scheduling around issues that arise in implementation"; "[give us] some guidance on how to divide work."

In Experimental Projects I and II, the teaching team often inquired about team skills and team process at the regular team meetings and also addressed team problems when they arose, but they did not lecture. When asked in interviews about what faculty could do to support team skills, students, agreeing with survey and focus group data, said that more lectures would not help. But at the end of the course, Peter said he thought that faculty could do a better job of making it clear that teamwork was a legitimate part of the technical project: "Just tell us about what could be a problem ... bring it out in the open ... a very helpful thing is talking to former teams not just about their experiment but how they worked together ... it makes people realize 'oh, this is part of it, it [teamwork] is a legitimate part of it so listen up!"

John, at the end of the course, suggested that emphasizing teamwork should begin much earlier in the curriculum:

> Starting teamwork earlier in the curriculum (not just the course) and making it an emphasis. Because I think the best [learning] is to be on a team and have the experience. Maybe working on teams earlier in your academic life. For me, this was the largest team project I had ever undertaken and I am a junior. [Faculty could] ... provide a list of previous students as a resource. [Faculty could] ... have a series of discussions about common or likely team issues as you go through the project. Then people can share different personal experiences.

Barbara thought that increased assessment of what was actually going on within the teams would help focus the faculty's efforts: "Perhaps doing some individual assessment or evaluation could be helpful ... let's say the faculty sends me a document with a few questions on it: is your partner doing her fair share of the work? What are your partner's strengths? What are your partner's weaknesses? Are there conflicts that you are not able to resolve? And then I could write back and my partner would never see my answers but the faculty could be aware of difficulties and get a little more involved."

Kaya, whose editorial experience put her in a leadership role with fifty-two student writers, said students would have complained about lectures on team skills: "They wouldn't see the big picture of why they were getting this lecture on day one when they could be in the lab talking about design. So you would get negative kickback from that. But if you did it at the end, people would say, 'Why didn't you tell us this on day one?'"

But when asked about more explicit structuring for the collaborative communication work, Kaya confirmed that explicit guidance and organization from the faculty would be more effective: "I think any form of structure from professors would be more well received than any structure from students ... some students would ask, 'Why are we doing it like this?' and if I said 'Because the faculty want it this way', then they're fine with that."

Students consistently reported that they did not want lectures from faculty, but just as consistently they mentioned faculty as mentors, and, in fact, they had expectations (albeit not clearly defined) about what that mentoring constituted. When mentoring was inadequate, they mentioned that too. Their responses revealed their intense awareness of what they thought they were or were not receiving.

In focus groups, some students commented appreciatively about their positive experience with mentoring, connecting strong mentoring with increased learning. For example, one student said in the focus group, "My mentor was a big part of my learning. He prompted me and then came and checked on me and asked me questions about the next step and after a while I started to expect it and prepare for it." Another student commented that she "liked it" when the professor "grilled" her. She said that she "appreciated that kind of rigor."

Other students thought that they had not received helpful mentoring. One student said, "There was a lack of personal mentoring for me. It did not help." Others described their mentors as "not involved," "not very connected to the project," or "too busy to put in much effort." Another student regretted that for him, "mentoring concentrated only on technical skills and not on learning and improving team skills." And again, a student noticed what other students received: "Some people got a lot of mentor support and interaction; others got very little. I got very little."

The influence of more advanced students was useful too, although students did not rely on this resource as they did on their professors. A student commented, "Watching seniors interact was a valuable experience for me. For example, observing how they handled fellow students with different skill sets so that everyone was useful."

Faculty members indicated on surveys that they thought mentoring was how students best learn team skills, but some faculty responded that they were constrained by time or by uncertainty about their role. One professor commented, "I do not see that [mentoring] as my role. There is not enough time for me to deal with [team process] too." Another faculty member wondered, "How much should I help them with team skills? How much should they figure out for themselves?" Another faculty member asserted, "Students do figure it out for themselves." An engineering professor described the faculty role as "mostly a role model and a consultant." Students learn by doing, he claims, and only when the team process is not working do they think about it. When the team process becomes too dysfunctional, his role, he says, is to "jump in with explicit advice and help them sort it out before they waste too much time" (E. Greitzer, personal communication, December 31, 2008).

One professor described a mentoring process that was highly individualistic. Based on her technical knowledge and extensive experience in her profession, she held her students to a high professional standard of technical and collaborative work. Yet she did not teach these skills separately or explicitly. At this level, technical and collaborative skills are completely integrated. "I am very adaptive," she explained. "I see what an individual student needs . . . where she or he needs to go and I push them—drag them!—toward those goals" (S. Widnall, personal communication, December 19, 2008). Using positive reinforcement or critique and steady feedback, she shaped students in their technical and collaborative development.

Clearly, *mentoring* has varying definitions among students and faculty. Some describe (or expect) a highly individualistic and personal relationship, while for other students and faculty, the relationship is more removed although perhaps still attentive. For some, the relationship is on an as-needed basis; for others, the relationship may be desultory. The actual focus of the exchange may be strictly on technical matters, or it may expand to include collaborative and professional skills as well. The mentor may be engineering faculty or adjunct faculty, an industry professional, a graduate student, or technical staff. The common characteristic described in this handful of interviews seems to be a person-to-person engagement between a novice and an expert (of one sort or another) in which the novice's activities are commented on and shaped by the more experienced mentor.

Students find specific verbal and written feedback helpful as they learn to communicate collaboratively.

Sometimes students lack a steady and positive mentor or role model, and although their collaborative experiences are extensive, they are not successful. Sometimes feedback to a student is not coherent or meaningful. How does a student in an unproductive or unmentored situation identify and then begin to improve weak collaborative skills and achieve a better result? Although assessment is often thought of as a way to compose a grade for a course, thoughtfully designed assessment, formative and summative, and reflection methods are useful guides for students.

In Experimental Projects I and II and in Space Systems Engineering, a good deal of effort goes into formative assessment of student work and team skills before the final evaluative grade is determined. Consequently students receive written and oral feedback on rough drafts, in writing conferences, and on presentation rehearsals. In Experimental Projects I and II, they receive feedback from the teaching team during team meetings. The feedback in these instances is on technical aspects of their work and on communication skills and, at times, on collaborative skills. They receive feedback in handwritten comments on hard copies, electronic comments on electronic copies, written comments on rubrics during presentations, and verbal feedback during team meetings and after presentations.

In Space Systems Engineering, students receive peer assessment grades and comments twice in each semester, and they receive mentor grades too. In addition, their team meetings with disciplinary mentors offer opportunities not only for technical consultation but also for feedback from their mentors.

When asked, a majority (77 percent) of students responded that dry run or rehearsal sessions for reviews were "useful" or "very useful." A slightly lower percentage (60 percent) responded that preliminary drafts of design documents and proposals were "useful" or "very useful." Lectures and seminars about collaborative communication strategies were not seen as very useful (50 percent). The usefulness of writing conferences varied depending on whether the students were in Experimental Projects I and II or Space Systems Engineering, with the Experimental Projects I and II students valuing those conferences more highly.

Grades for a collaborative communication assignment or for the course are meant to come after the preliminary and formative assessments, and these are often coupled with more written feedback. Different teaching teams devise their own grading methods for evaluating collaborative communication work. For example, in Experimental Projects I and II, the disciplinary professors, the graduate teaching assistants, and the communication instructor all provide grades, using rubrics to help ensure consistency. The students receive all the commented drafts, commented rubrics, and the grades for their communication deliverables, thus learning how different listeners or readers perceive their work. At the end of the term, these grades are combined into a collective grade, along with grades from the technical staff and the project advisor. Students are clearly informed about how this ultimate collective grade is composed. Space Systems Engineering faculty compose student grades in a similar fashion. However, in other courses, the disciplinary professors prefer to give only the collective grade. In these courses, while students may receive feedback (written or oral) from the various members of the teaching team, the team meets to arrive at a single collective grade.

In 2008, students who were completing or had just recently completed Experimental Projects I, Experimental Projects II, Space Systems Engineering, and Flight Vehicle Design (a third capstone option) were surveyed about collaborative communication (oral and written) and their thoughts about the assessment and grading. Because the courses differ in size and organization, not all students had experienced all modes of assessment and evaluation. For example, students in Experimental Projects I and II always had individual or team writing conferences, while only the editing team in Space Systems Engineering had similar conferences. Nevertheless, all students had experienced most of these assessment and evaluation strategies.

Responding to a question about multiple pieces of feedback and grades, a strong majority (68 percent) of the students reported that multiple technical grades were "useful" or "very useful" in helping them learn to write and present. In addition, a majority (68 percent) said that a communication grade separate from the technical grade was "useful" or "very useful" to them. An overwhelming majority of students found written feedback (91 percent) and verbal feedback (82 percent) to be "very useful." A student commented "Written feedback is the most useful because you can refer to it later, and when it is accompanied by a grade, it is especially helpful."

Moreover, just over half of the students claimed that the multiplicity of grades was "not difficult or confusing to understand" or rarely so. One student, clearly unperturbed, commented, "Different people know about different things." Another commented "It was helpful to have all of the faculty write comments on evaluation sheets for the presentations because different faculty identified different problems." Of the remainder, just over a third of the students found the multiple grades to be "sometimes difficult or confusing to understand."

Conversely, students did not appear to think that collective grades were useful to their learning. Students in the survey said they did not find a collective grade (a grade in which technical and communication scores from multiple instructors are combined) very useful because they could not extract enough specific information from the single grade. A student wrote, "The merged grades were useful only if there was a breakdown of how it was achieved—how much was due to technical work and how much due to communication work. Without that, the merged grade was useless." Another student wrote, "Would have been helpful to receive more comments. We learn by feedback." And just half of the students felt that the collective grade was "usually" or "always" fair to them.

When asked how they used the comments and grades that they and their team received on collaborative communication, a strong majority (85 percent) "often" or "always" tried to implement that feedback in later writing and speaking assignments. And over half (68 percent) of the students said that they "often" or "always" reflected on these comments individually. Less than half of the students said that they reflected on the feedback with their teammates.

What does one do with all that feedback with its multiple, helpful, and sometimes conflicting perspectives? Ideally, we hope that our students learn to reflect on their actions and abilities. Reflection—that ability to evaluate one's abilities—is part of meaningful behavioral change, but it is a practice that is difficult to encourage, hard to observe, and impossible to quantify. Perhaps reflection or self-evaluation often has to be prompted through an assessment tool or a team activity. After all, a graduate student commented, "People do not reflect when things are going well. It is when the feedback on the work is negative, when things get dysfunctional . . . then you begin to reflect on the way in which you and your team are working." Echoing this, Karen wrote in an e-mail sent after the research interviews were finished, "Our project went fine. I was always glad Charles was my partner . . . reflection was not necessary." Similarly, a student in Experimental Projects I who was working successfully with her partner said, "All this reflection stuff—seems a little silly. There is so much technical work we have to do."

In contrast, Barbara, who had worked successfully (and patiently) through several conflicts with her partner, wrote:

> When I was working with Amy, I never consciously thought about "teamwork." But the interviews helped me focus on different teams skills. So for example, after we talked about communication, then I got to thinking about it and then whenever I was working with Amy, I consciously tried to be more clear and make an extra effort to explain and then that helped dispel any tension between us. . . . I think the process of reflecting on teamwork helped to teach me more about team skills than any lecture or class.

Reflection is often mentioned as one of the higher-order thinking skills in critical thinking taxonomy, emerging as part of individual maturation (see Facione 2006, King and Kitchener 1994, Bloom and Krathwohl 1956, Perry 1970, Halpern 1998). In the process of reflection, the learner not only receives the feedback of others but also evaluates that feedback. The learner considers the perspectives of others, questions assumptions, and eventually synthesizes the information in a way that allows him or her to move forward with new or perhaps newly aware behavior and transfer that behavior to new situations. Yet maturation proceeds at its own pace, possibly explaining why some students resist well-intentioned guidance while others eagerly learn from it. However, our task in teaching students to develop professional communication skills is not to wait for maturation to do its work, but to find ways to meet students where they are in their development and encourage them to move forward.

Perhaps responses to the question of how students best learn collaborative skills can be distributed along a continuum. At one end, teams are working successfully on technical tasks, effectively summoning the necessary technical, collaborative, and management skills in an integrated flow. Some mentoring or positive reinforcement may be all that is needed to guide students to continued success. In fact, team success naturally reinforces team behaviors, so that strong, functional teams, once in motion, often continue to improve unless some unforeseen crisis occurs. Already rich in useful team skills and ready to learn more, teams working successfully may find blocks of information or structured methods of reflection to be beside the point—a bit fussy, not too useful.

At the other end of the continuum, teams struggle, fracture, and become despondent and lose motivation. The success of their experiment or their project can be affected. Between those points, teams work through various episodes of conflict, managing to mend their differences, solve problems, and fashion new skills. They often succeed, but it requires time—time that might better be devoted to their technical work. For these teams, some didactic material, focused assessment, skillful mentoring, and structured reflection may be the only ways to establish and strengthen the collaborative skills necessary to address their technical tasks and communicate their work collaboratively.

The focus groups, interviews, and surveys reveal that students are eager for feedback, and they take it seriously. When they are successful, feedback or assessment helps them see their strengths, and when they are not so successful, it helps them diagnose the sources of their difficulties. It appears that although students do not want a lecture on team skills, they do want to talk about collaboration. One student said, "Don't talk *at* us; talk with us [about collaborative skills]." For engineering and communication faculty working to help students succeed at complex technical work and collaborative communication, methods of mentoring, formative and summative assessment, and structured reflection are useful to some degree for all students who are learning to collaborate. And when student teams begin to struggle and even fail, the early implementation of those methods is especially important.

Summary of Writing and Speaking Collaboratively

The student and faculty voices that emerged in the focus groups, surveys, and interviews of this research illustrate what it means for students to learn to write and speak collaboratively in an engineering discipline. Our respondents confirmed that collaborative communication is as much about management and team skills as it is about the writing and speaking process. Moreover, the students told us that shifting from an individual process to a collaborative communication process was difficult for them and that we could help them by being explicit and structuring that transition more clearly. Students also indicated that we do not need to convince them of the importance of team skills, but they thought that faculty could do a better job of sharing their experience with students as they learn how and when to use those skills. Students and faculty alike thought that mentoring was a good way to share the knowledge of experienced professionals with students, but definitions and expectations of mentoring differed, and despite the varying definitions, this kind of teacher-to-student engagement was often constrained by demands on faculty time and perhaps by faculty uncertainty. Nevertheless, mentoring emerged as the method deemed most effective with advanced engineering and science students who are learning and refining their collaborative and communication skills.

In this conversation with our students, we remember Anson and Forsberg's (1990) description of this stage as a transition—much more than simply learning to write. Our students enter this transition with the expectation that what they know and the way they have always written will serve them in this new endeavor, and inevitably they are frustrated and confused when their student skills do not quite suffice in the more authentic writing tasks that we have set before them. They often distance themselves from us, feeling that as bright, competent students, perhaps they ought to know more or be able to figure this transition out. Their learning is not linear or predictable or general. Often it is on a continuum of maturation that is specific to each individual. Yet when their faculty can provide helpful structure and useful feedback—perhaps most desirable—the mentoring that students say they value, our students make this transition to a new professional identity: skillful collaborative writers and speakers with strong team skills.

6 Conclusions

Our intent in this book has been to add to the ongoing research in identifying best educational practices in communication-intensive (CI) classes. Our research has explored the discourse communities of biology, biological engineering, health sciences technology, biomedical engineering, and aeronautic and astronautic engineering. Each case study shows the rich, discipline-specific ways of knowing and communicating that face students as they learn to take on the identities of professionals in those fields. Despite the differences among these fields, common observations about CI pedagogy have emerged. In this chapter, we detail those findings in the hope that colleagues outside MIT can use these insights in the design of their own curricula.

We divide our discussion about the pedagogical implications of our research into several sections: the learner, the classroom setting and instructional method, the teacher and teaching team, assignment design, and assessment and reflection.

The Learner and Student Development

As the case studies in this book demonstrate, learning to write in the disciplines shares many of the developmental trends found in other writing research (see, e.g., Carroll 2002, Thaiss and Zawacki 2006). Learning to communicate in science and engineering classes does not occur as linear process or at a consistent pace (Sommers 2008). Some students showed remarkable gains by the end of the semester, while others were still struggling with core concepts even as the semester drew to an end. This is an important finding for faculty to consider; not all students will leave our classes having obtained the same skills. Learning goals for classes should be broad enough to accommodate the range of learning outcomes with which students leave our classes. Not only do some students make rapid strides while others make only small steps along their way, but students bring varied experiences, expectations, and attitudes to the task of writing in college. For some students, such as Jake in chapter 1 and Julia in chapter 4, that previous knowledge enabled them to go beyond some of the rudimentary expectations for their writing tasks. For other students, such as Barbara and Peter in chapter 5, the interpersonal skills that they brought with them enhanced their collaborative abilities, but even those skills developed further. Barbara and Peter learned to listen more closely and to be more patient. Still, for some students, such as Carla and April in chapter 1 and Nedra in chapter 2, the in-flux nature of their choice of majors and careers seemed to overwhelm the attention they could give to their writing tasks.

For the students we studied, prior professional and technical experiences influenced their development of communication competences, although they were not the most important factor. For instance, specific scientific experience could be helpful for some students, as it was for Jake in SciComm and Julia in Quantitative Physiology, but not all students could directly apply these experiences to the writing they were asked to do, as was the case for Kay in chapter 2. The more important factor seemed to be their perspectives on the rhetorical tasks that they faced, whether that task was a grant proposal or a scientific figure or a collaboratively written document. In other words, how well students managed their writing tasks was often influenced by what they knew about the rhetorical situation, how it was similar to and different from previous rhetorical situations they had encountered, and how well they could project a future identity onto the situation. For example, the students in chapter 3 knew little about the NIH grant writing process, yet they could draw on their previous writing experiences and shape that knowledge to the new writing demands of grants. Peter and Kaya in chapter 5 drew on their management and leadership skills from military training to help them organize their collaborative communication tasks.

Of course, few students come to our classes with advanced rhetorical perspectives (that is what they are there to learn, after all); what emerges as perhaps most important is for students to develop flexible rhetorical repertoires and be able to apply elements of previous communication experiences to new ones. To paraphrase David Bartholomae's (1986) notion of basic writing, that students need to "invent the university" in order to succeed with typical college writing tasks, students in our study needed to "invent the scientist or engineer"; how well they could do that depended on their relative levels of experience, maturity, and cognitive ability.

The cases in this book also attest to the need for a certain readiness to learn to write as a scientist or engineer. After all, Vygotsky's (1978) developmental theory of the zone of proximal development, or the idea that learners are capable of higher achievement if working alongside a more experienced peer or instructor, still implies differences in that learning and between what different students or novices are capable of alone. Baxter Magolda uses the term *self-authorship* to describe how students work toward "making meaning of world and oneself" (1999, p. 6). Key to this idea is that students move toward developing an idea of themselves as "capable of constructing knowledge, as in control of their thoughts and beliefs, and capable of expressing themselves" (Baxter Magolda 1999, pp. 253–254). In writing and speaking about the content of their technical disciplines, students move from merely doing homework to finding their young professional voices and perspectives. In finding their professional voices, we witnessed the ways in which they then could take on additional dimensions of professional identity. For example, in chapter 3, as students took on the identities of professional grant reviewers, they began to assume certain attitudes and perspectives toward scientific peer review more generally. That brought forth a change in how they saw the purpose of scientific grant writing. In chapter 5, students became insightful critics of oral presentations as they sat through the rehearsals of design reviews.

Yet some students are not ready, and they resist communication instruction and practice. For example, in collaborative work, not all students willingly worked in teams, and many students brought negative group experiences with them. In other settings, students were unenthusiastic about revising their work, especially when they received contradictory feedback from multiple reviewers—a new experience for many of them. A survey in the Department of Aeronautics and Astronautics documented that most students do not find multiple sets of feedback confusing; however, some students may need to be guided through the real-world experience of getting conflicting reviews.

In addition to observing differing developmental trajectories, the impact of prior knowledge, and students' readiness to learn, we noted the importance of stretching students' writing identities in several ways. Professional scientists and engineers do not write or speak in just one or two genres. They communicate in a multitude of forms to various audiences through various mediums at different rates. As shown in chapter 2, exposing students to these various forms of professional communication gives students a sense of the range of communication activities professionals use: poster presentation, lab talk, conference talk, "elevator" speech, grant pitch, interview, or job talk, for example. Just as important, students faced with this professional range can begin to envision the professional activities they themselves may undertake. Their development of a professional identity can be enhanced when they explore the boundaries of professional discourse as well as the well-worn paths as represented by some professional communication tasks. Bransford et al. (2000) describe this process as one of "adaptive expertise." Unlike routine experts who seek to solve limited problems efficiently, adaptive experts can apply existing knowledge to novel problems. Adaptive experts "are more likely to relish challenges that require them to 'stretch' their knowledge and abilities" (Bransford et al., 2001). An important finding here for faculty is that to develop students' professional adaptive expertise in communication, instruction needs to attend to more than just grammar or the routine elements of writing and speaking.

In the end, despite their individual developmental stages, all students involved in this research benefited from communication instruction and collaborative work. At times, they refined and strengthened what they already knew how to do, and at other times, they learned something entirely new.

The Influence and Context of Schooling

One aspect of each of the case studies we have described is perhaps obvious: most of the settings we studied are laboratories in which students perform hands-on science and engineering research or design. Such contexts for learning writing and speaking are not common. Often students learn about professional writing in classes that are divorced from the work of producing technical knowledge. In our findings, having students engage in the authentic work of science and engineering, including the communication tasks associated with that work, was an important condition for student learning. Communication was no longer orthogonal to scientific and technical content; it was integral to the process of making knowledge.

Yet we realize that the pursuit of authenticity is often elusive. Despite the faculty's care in modeling professional settings, the communication work still occurs within the context of school, introducing an inevitable and powerful element that makes communication like the work of professional scientists and engineers but not that actual work.

Studies of students learning to write in higher education have consistently shown the strong influence of schooling itself as a constraining factor (Carroll 2002, Dannels 2000, Haswell 1991, Sternglass 1997, Beaufort 2007). For our students, particularly Nedra in chapter 2, school was a powerful factor, whether that meant the time they decided to make available for their writing and revision or the structure and schedule (and potential restraints) of when the writing was due. These factors seem inevitable at most institutions given the time line of the semester or quarter and the need to assign grades.

Thus, schooling both enables learning in the direct instruction students receive in learning how to write and speak as scientists and engineers and constrains that learning in the limits of school-lab scientific results and the pressures of time management and grades. Students in our study did not dismiss the writing tasks they engaged in because they did not map directly on to post-college tasks; if anything, most recognized the utility of the tasks to the work they would do eventually, and in the best cases, they treated the constraints of schooling as part of the overall rhetorical situation. After all, writing in the science and engineering real world is by no means free from deadlines, navigating personal relationships with colleagues and coauthors, or evaluation by supervisors, coauthors, and clients (Winsor 2003). School can be an extremely useful space to learn professional communication precisely because it is a controlled space. In other words, slices of professional communication practices can be pulled out and taught in school contexts in ways that focus students on these core elements of professional communication. A professional scientist, for example, is unlikely to get a three-month course in making arguments with data, although it would be appropriate and feasible for a CI course to focus on this element of professional communication. And the authentic worlds of professionals are not always conducive to learning. The authentic work of science and engineering can sometimes be critical, arbitrary, hierarchical, and more about sustaining the status quo than innovation.

The pedagogical approaches we have described in this book may require a shift to a different style of classroom. For example, collaborative communication instruction based in active and cooperative learning methods results in a classroom that is likely to be noisier and less orderly than conventional lecture-based classrooms. Desks and chairs are dragged around, or students may need break-out space so that subsystem teams can work alone. Some teams may eschew the classroom altogether and head for a machine shop or computer cluster. In other cases, students may need access to a lab to gather more data outside official laboratory times. In all of these cases, as the activities become more authentic, the classroom becomes a workplace of sorts: busy, less predictable, goal oriented but process driven. The relative calm of a lecture-based classroom is exchanged for a quality that is more desirable: student engagement. While this last observation may be natural for faculty with a writing background, it is not as obvious for faculty who have always taught in lecture halls or have tightly controlled lab times. Conversely, for someone with a writing background, this insight means looking for alternative spaces to integrate instruction, such as the laboratory or the field—spaces where students often have time for instruction while they wait for an experiment to run. It also means employing digital media in the delivery of instruction outside class time. As the classroom becomes more like the workplace, deadlines, meetings, or work sessions can fall outside scheduled class times.

In the end, we believe the benefits of flexible learning environments far outweigh concerns about artificiality, and the effort to create these contexts is well worthwhile. Experiential learning brings to bear the sociocognitive perspectives we have described throughout this book, and it is difficult to imagine any useful learning that would stem from having students working solely from textbook examples. In a sense, in designing the settings for communication instruction, we might refer to the long-standing tradition of fieldwork in science instruction (Gladfelter 2002) and imagine the many ways the "field" might be represented to our students or the ways that their writing and speaking tasks might be matched to the exigency of real audiences, as was true in several of the classes we described.

The Teacher or Teaching Team

The role of the teacher (or, as is often the case at MIT, the teaching team) in the kind of communication pedagogies we have described usually differs from the traditional educator role. Engineering and science faculty in this research described their roles using various terms: *mentor, consultant, coach, role model,* and *advisor.* Whatever the term, the common characteristic was that of active engagement in the educational setting rather than a removed and formal presence. Faculty in an engaged classroom do not simply lecture about the concepts of professional communication, persuasion, or teamwork; they may also take an active and visible part in the discussion. For students who may be ambivalent about the importance of communication or collaboration strategies, the faculty member's modeling and affirmation of those skills are critically important.

We also recognize, however, that faculty need support in designing engaged classrooms, especially if they have little experience teaching writing and speaking. Many universities and colleges offer workshops on best teaching practices in writing and speaking. While workshops on integrating writing in disciplinary classes often focus on assignment design and assessment, they tend to be general assertions of best pedagogical practices (*An Introduction to Writing Across the Curriculum* 1997). In the classes that we studied for this project, faculty often had long-standing collaborations with communication instructors from the MIT WAC Program, who themselves were familiar with disciplinary writing. Over time, however, the communication instructors developed a deeper sense of the disciplinary and classroom demands in each context, which allowed them to suggest further ways of making classroom activities more attuned to the goals of authentic learning. Such collaborations allowed the communication instructors to provide repeated support for faculty, which was especially important after failed pedagogical attempts. Most important, through those collaborations, faculty were encouraged to tap into their deep disciplinary expertise in scientific and engineering communication to make that knowledge more transparent for students. We recommend such consistent collaborations, even if the contact must be limited because of time or resources. In our experience, inserting writing instructors into classes where they only grade papers or conduct writing workshops that are disconnected from the course content is not successful or efficient.

For some faculty, such as Natalie Kuldell (chapter 2), Sangeeta Bhatia (chapter 3), and Sheila Widnall (chapter 5), the shift in their roles from teacher to coach or mentor seemed to come more easily than for other faculty. Perhaps one reason for their enthusiasm was their sense that they could make a connection between their mentoring role in the classroom and the professional mentoring roles that they play with their graduate and postdoctoral students. For others, the hands-on nature of communication instruction is a welcome change from the conventional expectation that the faculty member's job is to dispense knowledge and then judge students' mastery of that knowledge on multiple-choice exams. Through this expanded role, faculty experts make their tacit knowledge explicit for novice learners. Often we observed faculty members teaching students about how an audience receives a written document or review. Also, we have seen the ways in which engaging in communication instruction offers faculty and graduate students opportunities to reflect on their own writing histories and processes. Many of them welcome these opportunities.

When engineering and science faculty make their tacit knowledge known, it is worthwhile to present that knowledge as standards widely held by the professional community rather than idiosyncratic beliefs from one instructor. Some of the students we interviewed expressed some question as to whether they were responding to the specific demands of their instructor or to larger professional standards. In a sense, some students treated any class as a unique rhetorical context (i.e., they were writing for that particular instructor with little belief that their instructor might represent some larger group). Perhaps this phenomenon is simply the result of disciplinary knowledge broken into discrete blocks seemingly without connection to one another. However, it does indicate the need for professionals to stress both the generalized nature of their standards and the sources for those standards, as well as the potential for students to contribute to the development of standards and "own" the criteria on which their writing is judged (see, e.g., Inoue 2005).

Finally, communication instructors working in disciplinary classes face their own challenges. Initially a communication instructor working in a class such as Space Systems Engineering might feel as if she had landed on Mars, encountering technical content so impervious that it is difficult to imagine teaching beyond the fundamentals of technical communication. Yet as we begin to observe how faculty and students talk and write, not just what they talk and write about, we can begin to observe what faculty as professionals are doing, expecting, and seeing and how that activity may differ from what students are doing, expecting, and seeing. Well-placed communication instruction can help bridge some of those gaps and get students moving toward the goal of writing and speaking as professionals. At times communication instructors might be challenged for not being aeronautic and astronautic engineers or biomedical engineers, but what communication instructors bring to the science and engineering classroom is expertise in engineering and scientific communication and knowledge of how to use communication instruction to promote learning. In turn, our engineering and science colleagues rediscover and affirm what they have always known and practiced: the absolute necessity of clear communication in their disciplines. As they mentor students, our colleagues become strong advocates for writing and speaking.

Communication Assignments

A long-standing element of engineering and science education is to offer specific tasks that have practical ends, and we certainly see in our students and our alumni a strong desire to learn specific communication skills that will be of immediate use when they are on the job. While we acknowledge that the match between classroom and on-the-job learning is uneven (see, e.g., Artemeva 2005, Beaufort 2007, Leydens 2008), ignoring these demands seems misguided and runs the danger of making students doubt the value of communication instruction. The tasks from which students learn to write and speak as scientists and engineers need to be designed with students' long-term careers in mind, whether that translates to the specific technologies they will be using or the foundational knowledge they need to possess.

One significant complication to crafting the connection between communication tasks learned in school and postgraduation realities is that learning that is the result of these authentic tasks is not always what is expected or planned. For some of our participants, the outcome of their writing experiences was to change their majors. For others, the outcome was to extract the skills and strategies they would need for the long term. For any student, the outcome might not necessarily be the mastery of specific tasks; rather, they are exposed to the range of discursive activities that professionals engage in and the ways that professionals attempt to make meaning by writing and speaking. As we pointed out earlier, students come to us at different points on the continuum of learning and progress at individual rates.

Thus, an implication for pedagogy is that assignments have to be flexible enough to accommodate students at their various stages of development and that our assessment of the success or failure of those assignments is not necessarily based on whether students receive a high mark on the final product. For example, the success of a particular assignment might be that it exposed students to the real work of a biological engineer, and they discovered that they did not like it and switched to a more suitable major. The parallel here is to the work of experimental science when failed experiments are the norm, providing valuable knowledge about the next experiment to conduct. The productivity of failure, however, is not easily accounted for in standard assessment of student learning.

Another difficulty for assessing student learning is when outcomes are not aligned with initial goals or will not become clear until long after the class is over. Can such long-range effects be considered in the short-term planning needed for assignment design and sequence? Certainly scientists and engineers faced with communication tasks might not immediately reflect on their student experiences, a finding confirmed by Leydens (2008), who found that two years into the job of an engineer, one participant had no memory of writing the same task as a student that he now faced as a professional. However, the same charge could be made against a good deal of what students learn: that the mapping on to future experience is never entirely direct. Writing classes have perhaps been unfairly criticized for this inevitable abstraction or lack of transfer of specific skills, but once again, recognizing the larger goals of professional preparation is key here. Students will learn to write as scientists and engineers by engaging in professional tasks. That writing, of course, will inevitably be professional-like, a novice approximation of the skills they will bring to bear on these tasks once they have the full benefit of experience and additional instruction.

Students might bristle at this degree of artificiality, as was true for some students in our studies, given the tension between the quality of their research results and students' realization that in the real world, what they had produced in the lab was not publishable. Nevertheless, it is important again to note that the writing activities were apprenticeship activities, offering much of the intellectual rigor and challenge of real-world tasks but with support structures built in. Given students' experiences in these classes, one implication is that the ways that tasks are not authentic need to be acknowledged and the larger learning goals articulated. For example, in Quantitative Physiology (chapter 4), the faculty openly acknowledged that students were not gathering enough data to sustain a publishable research article. And in Frontiers (chapter 3), Bhatia acknowledged that first-year graduate students were unlikely to have sufficient preliminary data for an NIH R01 grant. Yet these limitations did not impede the aspects of the learning experience that more closely approximated authentic activities of professional practice.

We have found that many standard technical writing assignments do not serve the purpose of teaching students effective communication skills because they neither approach authenticity nor offer opportunities for practice and revision. For example, the common final report in which pairs or teams of students collaborate on a final document and then submit it in the final week of class without any interim feedback or assessment seems like a worthy effort, but such an assignment is virtually worthless in regard to learning to write. Similarly, an oral technical presentation without rehearsal usually produces a group effort in which students deliver an unsatisfactory first draft of their presentation, receive average or poor grades, and are likely not to have another chance to improve their performance. Opportunities for learning to write and present at a professional level are sacrificed when assignments are not appropriately chosen or developed.

Ideally, productive writing or speaking assignments are divided into phases or stages in which individual students or teams have the opportunity to work on smaller, less complex pieces and receive feedback. Fortunately, in engineering education, many student projects are easily divided into stages as students develop an idea from conception to operation or testing. Likewise, in science, the research process can include proposal writing before entering the lab to collect data. The intelligent integration of communication practice into these naturally occurring stages builds students' sense of professional community and provides meaningful content for subsequent communication tasks. Disciplinary faculty members are a valuable resource in determining where to divide tasks if we encourage them to make those decisions based on their professional values, standards, and genres.

Yet both communication instructors and disciplinary faculty need to develop communication assignments that are properly scoped and scheduled. Assignments that are too small and inauspiciously scheduled can discourage and disinterest students as much as assignments that are too large or too time intensive. Assignments for which students may not have the necessary skill sets can be similarly unsuccessful. For example, in collaborative settings, unstructured communication practice is usually not efficient or effective for the students who usually end up struggling (unless there is a supremely talented student on the team), performing at a barely satisfactory level, practicing not success but instead a series of failures, and ending by disliking communication tasks. Explicit structuring of the writing process increases group awareness and coordination and helps teams of writers produce writing or presentations efficiently and effectively (Lowry, Nunamaker, Curtis, and Lowry 2005; Chisholm 1990). Of course, structure in CI pedagogy is ultimately to be reduced as student writers and speakers absorb professional practices and go on to use and reuse that knowledge, making it their own. The objective is to have students achieve a high level of ability and then transfer that ability to other, perhaps unforeseen, contexts. Skillful teachers pay attention to the reduction of structure, so that their students will gain control of the skills.

Assessment of Student Performance

Assessment of student performance often conveys one image: grades. But assessment can take many forms, including formative assessment, peer feedback, summative evaluation, and the often overlooked self-assessment that comes from reflection. Often in the classes we studied, a mix of assessment methodologies was used: communication assignments, peer feedback, conferences, quizzes, and exams. Regardless of the assessment tool used, effective formative and summative assessment ultimately begins with well-designed assignments that are linked to clear learning goals (Yancey and Huot 1997). Clear learning goals help frame, not constrain, what students should take from a class.

In the classes described in this book, formative assessment was the most common approach to the assessment of student writing and speaking. For example, the initial drafts that students produced were commented on extensively by engineering and science faculty, as well as by communication instructors. In some classes, students received a grade for their rough draft to encourage them to produce a complete document at this initial stage, but this grade was usually a limited range, such as "excellent—acceptable as final paper," "significant work but not acceptable as final paper," or "not acceptable as draft."

Faculty feedback on drafts was decidedly formative and modeled a range of professional competencies. We call this *authentic feedback* because it, in part, embodies the evaluative standards of professionals. Authentic feedback, such as shown in chapters 3 and 4, pushed students beyond rote revisions of their writing to consider the rhetorical dimensions of their work. Were they really motivating their work convincingly? Were they making a convincing argument with their data? Authentic feedback for the students profiled in chapter 5 was often written on hard copy drafts or rubrics but also often was given verbally. As faculty feedback evidences in those chapters, authentic feedback need not be gentle, but it does need to attend to the big picture. Faculty who used authentic feedback rarely commented at length on low-level details and rarely copyedited student work.

Moreover, when disciplinary faculty responded as expert readers, they drew on the values and expectations of the professional community in which they are members. In other words, faculty supplied a professional and a disciplinary reading of and response to student work, quite explicitly teaching students what professionals value, whether on the level of language or in terms of larger units of discourse. Key to this process is that students then have the opportunity to revise their work and attempt to enact that feedback. As we discussed earlier, when faculty gave formative feedback that models professional discursive norms, students began to be immersed in those authentic assessment practices that are found in the professions.

Students also received feedback from communication instructors. Communication instructors responded rhetorically to student writing and oral presentations, inquiring about audience and argument, while also noting genre or conventions. Feedback ranged from rhetorical and organization features to specific grammar and language use. Graphics were similarly assessed. Communication instructors offered feedback in writing as well as in conferences with students. In our classes, we developed and shared our communications rubrics with disciplinary colleagues. From a larger perspective, one of our goals was to encourage faculty to respond consistently and meaningfully so that multiple reviewers reinforced feedback on communication assignments.

In addition to faculty and communication instructor feedback, students received feedback from peers, adding a real-world element that mirrors professionals' knowledge creation (Bazerman 1988). In chapter 3, we called this *authentic peer review* because it puts students in the role of professional evaluator. Peer evaluation would determine, for example, if a grant would receive funding. Peer review modeled on professional practice helped students move beyond many of the problems inherent in peer review, including uncertainty with assessment criteria and inexperience with giving and receiving feedback (Elbow and Belanoff 2000).

In addition to formative assessment, classes used summative assessment. However, summative assessment of communication tasks was always based on students' revised manuscripts and presentations, not averaged across drafts. And even when the final grade was given, even more feedback usually accompanied it. Moreover, most disciplinary faculty worked hard at aligning the criteria for assessment across various deliverables as students moved from draft to final version and then onto the next series of assignments. For example, in Experimental Projects I and II, the engineering faculty members worked to refine rubrics (presentation, proposal, final report) that made assessment criteria consistent throughout the two-course sequence. A similar effort was made to make criteria for oral presentations consistent too. Thus, in addition to helping achieve consistency among evaluators and over time, the rubrics reinforced the professional standards for students.

Keeping a consistent message about core criteria for success, however, can be challenging when class documents are edited by multiple faculty and spread over class management sites, the Web, lecture notes, and paper handouts. Each time a course is taught by a new faculty member, those criteria for assessment have to be revisited and some process put in place to ensure that the supporting documentation for the course retains a consistent language about assessment standards.

Despite the value of formative assessment, academia often seems to revolve around the course grade, and perhaps this is only natural since the grade point average is the gate to graduation and, possibly, future success. In individual work, summative assessment is straightforward, yet in collaborative work, summative assessment can be challenging. Some educators argue that collective work should not receive individual grades for individual portions of the work since that individualization destroys the cooperative spirit (Pappas and Hendricks 2000). Others argue that one holistic grade for the entire team motivates all students on a team to contribute their best (E. Greitzer, personal communication, June 11, 2008). However, some students find it unfair that the same grade is given to hard-working students as to their less productive peers. The literature on teamwork documents a range of solutions to the problem. For example, some portion of the grade can be tied to individual writing. An exam for each individual can check and confirm individual understanding. Peer-based evaluations can individualize student contributions to a project (Chisholm 1990; Smith, Sheppard, Johnson, and Johnson 1995; Kaufman, Felder, and Fuller 2000; McGourty and De Meuse 2001; Loughry, Ohland, and Moore 2007). In the end, the debate about individual assessment versus collective assessment may turn on the objective of the particular course or assignment. If one of the course objectives is to replicate as authentic a professional experience as possible, then perhaps the professor may lean toward collective assessment. But if an objective of the course is for the student to learn a specific skill, then perhaps individual assessment can be more heavily weighted.

Beyond formative and summative assessment is one final assessment approach that is often overlooked. Reflection, an individual practice observed in some of the classes that we studied, is useful in the individual growth and development not only of students but also of faculty. Reflection is often braided together with assessment. For example, when faculty members articulate standards for professional assessment of communication tasks, genres, and skills, translate those into terms that students can grasp, and then enact those standards in their responses, they come to a greater awareness of their own understanding of professional discourse. This greater awareness can be clearly articulated to their students, and as in the Department of Aeronautics and Astronautics, the process also leads to further reflection on continuous improvement measures for pedagogy. For students, the metacognitive process of turning inward to integrate information about one's performance is a skill that many students already have and one that is worth supporting. Although this process requires little in the way of forms, technology, databases, or even paper and pencil, the ability to reflect leads to the ability to manage, shape, and strengthen our actions.

Program Assessment and Development

In this book, we have explored at length faculty feedback and other forms of formative assessment, but our research also speaks to programmatic assessment issues. Certainly these are concerns of any WAC program, especially in this age of "value-added" education.

In thinking about how the classes profiled in this book fit within our overall WAC program at MIT, it is worth mentioning that the courses we describe are just a few of the CI classes at MIT. The successful courses (and not all are successful) serve as best practice models for other CI classes in that they demonstrate the varied ways that communication instruction can be integrated with technical content. In one case, SciComm, integration meant a tutorial in which students met once a week with a communication instructor to discuss their research writing. In Quantitative Physiology, our communication curriculum was altered in order to focus on one aspect of professional discourse: how to make arguments with data in different genres. In yet other cases, Experimental Projects I and II, one CI class developed a structure that then mapped onto a following CI class, so that the communication curriculum could extend over two semesters. We see this variety of models for CI instruction as a strength of MIT's WAC program, and we assess the program's success or failure in part on our ability to meet the evolving needs of disciplinary faculty through innovative instructional models.

Skillful and focused assessment and reflection lead to individual development for both students and faculty, and the same rewards can accrue to programs. Because our CI work is integrated into technical curriculum in such a variety of ways, we have been reluctant to use a standard summative assessment tool to evaluate our work at an institute. Nevertheless, a recent MIT survey collected student and faculty impressions of the communication requirement, issuing a report in spring 2008 that found that "MIT students and faculty place a high value on communication skills and expertise, and students recognize their [communication requirement] experiences as contributing to the development of these skills" (MIT 2008, p. 2). While we were gratified to see these results, we also feel that ongoing assessment is needed to uncover not only student and faculty impressions of effectiveness, but more specifically how students were learning to communicate as scientists and engineers.

The research in this book is the result of our intention to look more closely at that learning process. We believe that teacher-based research in specific disciplines is the kind of assessment that can provide useful insights into which pedagogical practices improve teaching and result in increased learning. In this research, we have listened to our students and our colleagues while steadily observing the dynamics of our own classrooms. We feel certain that the results of our research will enrich our future teaching and our collaborations. We invite our writing colleagues to join us in this endeavor by investigating and assessing their own writing programs.

Communication instruction within the context of fast-paced engineering and science education will never be simple. For communication instructors, it means being able to articulate and practice the fundamentals of our profession while working closely within the boundaries of a technical discipline. The stakes are high for the engineering and science faculty as well as for the students. However, helping students meet the target competencies of professional communication practice is both necessary and worthwhile. We believe that our examples set forth here speak to the possibilities and the challenges in meeting those needs both inside and outside MIT.

A Last Word: What It Means to Communicate as a Scientist and Engineer

This book is an outgrowth of our teaching and collaborations with technical faculty at MIT. For more than ten years, many of our science and engineering colleagues have welcomed us into their classrooms and included communication instruction and practice in their teaching. Together we have developed and changed assignments, altered grading methods, and addressed fundamental questions about student learning. They are our colleagues, friends, and collaborators. In turn, we have come to tailor our communication instruction to the values and norms of professionals working in the fields of science and engineering. And while learning their professional ways of working, we have learned something about science and engineering, conducted lab experiments, and read their professional journals. Such reciprocity is rare in academia, where faculty often stay in their respective disciplinary territories and only occasionally venture to the other side of campus for day-long teaching workshops. At times, our collaborations have been frustrating, as our values and background knowledge seem so disparate. Yet with each iteration of our work, we have come to better understand and appreciate the varied perspectives we both bring to student learning. Our collaborations have led us to insights about how students learn to be professional communicators, insights that are the substance of this book.

MIT has made a strong commitment to supporting our WAC program. Yet our work takes place in the real world of faculty and student life. MIT students and faculty are not unlike their peers at other institutions when it comes to implementing a WAC program. MIT students sometimes skip class, they resist our feedback, or they do not seek help when they need it. Faculty are not always willing to seek our help in designing assignments, responding to student writing, and trying to manage integrating technical content with communication instruction. These pockets of resistance indicate to us the difficulty of the tasks that we have undertaken. Communication-intensive pedagogy, especially when it is coupled with science and engineering instruction, requires a good deal of forethought, strong collaborative relationships, and continuing assessment and revision.

It is also true that MIT is a resource-rich institution that has dedicated a significant number of lecturers to this endeavor. However, many other institutions already have or are instituting WAC programs (Bazerman et al. 2005). We have colleagues in a variety of these settings who scale work similar to ours to fit the size of their institution, their budgets, and their missions. Our colleagues see, as do we, the value of this work.

As American higher education struggles with demonstrating student outcomes in an era of accountability to potential "customers," students' development as writers and speakers is more essential than ever before. Students' literacy practices have always possessed a particularly strong cultural capital, whether it is a focus on what students do not know (e.g., "Johnny can't read") or a set of standards for what it means to "be educated." Our research in science and engineering contexts supports a somewhat less charged but no less important conclusion: all students, whatever the institution, need to write and speak effectively in order to excel as scientists, engineers, physicians, or entrepreneurs. Reaching that outcome has long been a concern of the WAC movement in terms of identifying best practices in the teaching of writing in the disciplines (Anson 2002). For our students and colleagues in science and engineering disciplines, our hope is that the case studies in this book begin and sustain a widening circle of conversations, research projects, and instructional innovations that will contribute to what it means to learn to communicate as a scientist and engineer.

Appendix A: Data Collection Methods

All participants in the studies reported in this book gave informed consent in accordance with MIT's Committee on the Use of Humans as Experimental Subjects, and pseudonyms are used for student participants to ensure their privacy. Participants were offered the opportunity to read and check earlier versions of this work for accuracy of the descriptions and soundness of the conclusions.

Chapter 1: First Steps in Writing a Scientific Identity

Student and instructor experiences in SciComm were captured using a mix of whole-class surveys, one-to-one interviews with target students and their SciComm instructor, and analysis of students' writing.

Surveys

At the start of the semester and during an Experimental Biology lecture, all students were asked to complete a brief, short-answer, pen-and-pencil survey on their experiences as scientific writers, their expectations for writing in Experimental Biology, the relationship between science and communicating that science, and the connection they saw between communication proficiency and their future careers. They were also asked to respond to a series of statements about what they saw as the role and importance of various sections of the scientific research article, which students rated on a Likert scale of 1 to 5 (see appendix B for the complete survey). Of ninety-one students at the start of the term, seventy-three, or 80 percent, completed surveys. At the end of the term during an Experimental Biology lecture, students were asked to complete a brief pen-and-pencil survey of short-answer questions on what they thought they had learned as a result of their SciComm experiences. They also were asked to fill out the same Likert scale on their knowledge of the features of and importance of a research article as they completed in the first survey. Of the eighty-four students who completed the term, forty-eight, or 57 percent, submitted final surveys. Students to contact for follow-up interviews were culled from the twenty-four (from two SciComm sections) who were taught by one instructor, Marilee Ogren, to control for instructor effects (SciComm was taught by four instructors in six sections during the spring 2008 semester).

Interviews

Four students were chosen and agreed to follow-up interviews; they received a small payment once the interviews were completed. The choice of students to interview was based on the variety of answers they gave to survey questions, particularly a range of backgrounds with scientific writing or writing in general and an articulation of the role of writing to their futures as professionals. Each of these four students was then interviewed face-to-face at the start and at the end of the spring 2008 semester. These interviews were recorded, transcribed, and coded with the use of the software tool HyperResearch.

Students also submitted all drafts of their SciComm research articles. The SciComm instructor, Marilee Ogren, was interviewed at the end of the spring 2008 semester in terms of her experiences as a scientist and scientific writer and her reflections on the writing assignments students completed in Experimental Biology SciComm.

Chapter 2: Taking On the Identity of a Professional Researcher

Similar to SciComm, student and instructor experiences in BE Laboratory were captured through whole-class surveys, one-to-one interviews with target students and instructional staff, and analysis of assignment descriptions and students' writing.

Surveys

At the start of the semester, all students were asked to complete a brief, open-ended online survey on their experiences as scientific writers, their expectations for writing in BE Laboratory, and the connection they saw between communication proficiency and their future careers (see appendix B for the complete survey). Fourteen students (out of twenty-two) returned completed surveys.

Interviews

From the fourteen survey respondents, four were chosen and agreed to follow-up interviews (and received a small payment after completing those interviews). Choice of follow-up participants was based primarily on the variety of answers they gave to survey questions, particularly a range of backgrounds with scientific writing or writing in general and an articulation of the role of writing to biological engineering professionals that was particularly striking. Each of these four students was then interviewed face-to-face-at the beginning and end of the fall 2007 semester. These interviews were recorded, transcribed, and coded with the use of the software tool HyperResearch. Student participants were also asked to respond to a brief e-mailed follow-up survey at the end of the spring 2008 semester (see appendix B). One student did not participate in a follow-up e-mail interview or submit written materials and was not reported on in chapter 2.

Writing Samples and Faculty Interviews

Students submitted drafts of the major writing assignments. Faculty and instructional staff from biological engineering with responsibility for this class, Drew Endy and Nathalie Kuldell, were interviewed during the previous semester (spring 2007) as part of a separate research project, and Kuldell was interviewed at the beginning and end of the fall 2007 semester in terms of her experiences as a scientist and scientific writer and her motivations for and reflections on the writing assignments that she created for BE Laboratory

Chapter 3: Carving Out a Research Niche

In Frontiers in (bio)Medical Engineering and Physics, our research goal was twofold: (1) to understand how and if students were learning the concepts presented in the workshops on rhetorical strategies in grant writing and (2) to explore if changes in their thinking about scientific persuasion might also be linked to changes in their evolving identities as professional scientists. To get at these issues, we gave students a pre/post survey, interviewed students two times during the semester, talked with students one time after the semester, and analyzed changes in their writing across drafts. We also interviewed faculty and TAs.

Surveys

At the beginning and the end of the semester, all Frontiers students were asked to complete a brief, short-answer, pen-and-pencil survey (a Likert scale of 1 to 7 was used) on their experiences as scientific writers, the structure of a National Institutes of Health (NIH) grant, and the NIH review process. All twelve students at the start of the term completed surveys. At the end of the term, students were again asked to complete a brief pen-and-pencil survey of short-answer questions on the same questions about the structure of a grant and the NIH review process. Of the twelve students who completed the term, eight, or 67 percent, submitted final surveys.

Interviews

Four students were chosen and agreed to follow-up interviews; they received a small payment once interviews were completed. Choice of students to interview was based on the variety of answers they gave to survey questions. Each of these four students was then interviewed face-to-face at the start and end of the spring 2008 semester. These interviews were recorded, transcribed, and coded. Interview questions explored their writing strategies for each section of the grant, the review process, their relationship with their mentor, and general impressions of the course experience. (See appendix B.)

Analysis of Student Writing

Student writing was analyzed for changes across drafts. In reviewing student written work and oral presentations, we used this commentary along with reviewer feedback to track changes in student writing. We also looked for evidence of teaching tips from class workshops in student writing.

Chapter 4: Learning to Argue with Data

In studying Quantitative Physiology, our goal was to understand what students were learning about making scientifically sound arguments using data. While we had conducted a pilot study in 2004 and each year collected various assessment data, we wanted to undertake a deeper analysis of student learning with the hopes of improving the course pedagogy.

As with the other studies in this collection, we used a three-part approach: (1) pre/post surveys, (2) two semester interviews and one follow-up interview, and (3) analysis of student writing and presentations. Surveys and interviews were conducted in fall 2007 with the exception of the follow-up interviews, which were conducted in summer 2008. Faculty and TAs were also interviewed.

Surveys

Surveys were designed to assess student preconceptions about the scientific research article or presentations that might conflict with the storyboarding approach. For example, if students believed that they had to present all of their research data in the results section, then we needed to address that misconception before the story-

Table A.1
Taxonomy of survey questions

Question type	Purpose
Background experience	Determine scientific writing experience
Structure of scientific article or presentation	Identify misconceptions about scientific research articles
Role of argument or persuasion in scientific writing	Identify preconceptions about objectivity in scientific practice

boarding approach would make sense to them. Likewise, we wanted to know if students had other extensive experiences writing scientific papers. Finally, we wanted to determine their confidence with completing original research, and their sense of the responsibilities of scientific readers. This last point was important in giving students peer review guidance. Table A.1 shows the types of questions and their purpose. (For a complete list of survey questions, see appendix B.)

All thirty-one students in Quantitative Physiology were surveyed. Of these students, seventeen returned the survey at the beginning of the semester, for a response rate of 55 percent. By the end of the semester, these seventeen students were surveyed for the postsurvey. However, three of these students had dropped the course, so we concluded with fourteen postsurveys.

Interviews

Based on their survey findings, we selected five students, each with different experience levels. Interview questions followed from five main topics: project description and rationale for the project, results, analysis of results, use of storyboarding, and role of feedback and revision. Student responses were thematically coded into the following categories:

- What was done and why that topic was selected
- What was found
- Why they decided to focus on those data and not others
- How they explained those data
- How they chose to present those data
- Impact of peer review and feedback

After sorting interview responses into these categories, we looked for explicit and implicit evidence that students used the storyboarding method as well as changes in their talking about data. These subtle changes in language, such as noting the effect of a visual on a reader or selecting one data set over another, could be evidence that our teaching was having an impact on student learning even though students may not be using the prescribed storyboarding method. By looking at these smaller changes in student talk, we could see if their way of understanding scientific practice in the processing of research data was changing: What were their stances toward research, reporting, analyzing, making it fit, and thinking about the audience? Were there changes in the ways that they thought about the reception of their research?

Analysis of Student Writing and Presentations

During the interviews, we used discourse-based interview techniques in which students were asked to comment on their written work or oral presentation. With a copy of the document in front of them, students could comment on specific graphs or revisions. In reviewing student written work and oral presentations, we used this commentary along with reviewer feedback to track changes in student writing.

Chapter 5: Writing and Speaking Collaboratively

This research began with focus groups, used several surveys, and then followed up by interviewing individual students at several points in their courses, Experimental Projects I and II. The various methods produced different sorts of data. The focus groups helped us understand what students did not consider helpful and, conversely, what they thought would be useful in helping them learn team skills.

Surveys

In surveys, we explored what students' attitudes about team skills, and we coupled that with data from their project advisors and the course faculty. We saw some interesting parallels between the project advisors and the postcourse survey results. But again, data from a short survey and a small population may not support major curricular changes. The last survey collected data on student opinion about collaborative communication practice, the method of teaching it, and the methods of assessing the collaborative deliverables.

Interviews

The individual interviews were by far the most data-rich explorations and provided an interesting comparison against the data from the surveys. The interviews also gave us a sense of how individual students develop over an eight-month intensive team experience. In this dialogic exchange and in the presence of authentic student voices, we garnered a stronger sense of what it means to them to become a skillful, functional team member and to produce high-quality collaborative work.

Focus Group

In spring 2007, three exploratory focus groups were held with the students in Space Systems Engineering. The students had spent three semesters (spring 2006, fall 2006, spring 2007) designing a rover vehicle that could survive a landing on Mars, navigate rough terrain, and accomplish high-level scientific experiments with or without direct human guidance. The focus groups were launched in response to the faculty's perception that the teamwork element of the course had not run as smoothly as could have been wished.

Students were invited to come to one of three noontime groups. When they arrived, they were invited to have lunch, fill out a survey, and sign a consent form. After a half-hour, the focus group discussion began. The students were joined by one of the engineering faculty; the communication instructor facilitated, and a second communication instructor took notes.

Eighteen students took part in this focus group, although only twelve completed surveys. These students were the ones who had made the commitment to stay for the entire three semesters of the course while others who began the course with them had either graduated or had decided to switch to Experimental Project I and II.

Student Survey

For fall 2007–spring 2008, a precourse survey was administered to students beginning Experimental Projects I and a postcourse survey was administered to those completing Experimental Projects II. This survey, less personally oriented, asked students to rank team skills important to the engineering profession in general and team skills that had been important to them thus far. The survey also explored students' opinions about the role of mentoring in their development as an engineer. Twenty-seven students participated in this survey.

Student Interviews

In parallel to the fall 2007–spring 2008 survey, interviews were held with four students: two juniors and two graduating seniors and two women and two men. The interview participants were chosen randomly from a small group of volunteers. None of the students were known to the interviewer. Each student was paid a small stipend for completing a series of four interviews.

The interviews took place in September 2007, December 2007, February 2008, and May 2008. The objective was to promote reflection at critical points in a student's Experimental Projects I and II course and capture those insights. The interviews were guided by but not limited to interview questions. Each interview was recorded with the permission of the student and transcribed.

An additional interview was held with a student from Space Systems Engineering. Although this course was not the site of this research, the student's current experience with collaborative writing was worth documenting. Thus, with the student's permission, the interview was recorded and transcribed.

Faculty Survey

In fall 2007, the group of faculty working with the students who were entering Experimental Projects I were surveyed. This group included not only engineering faculty but also project advisors, the teaching fellow, and the technical staff and mentors. This survey paralleled the survey given to the students, asking faculty to respond to questions about team skills in engineering, team skills in their own careers, and their perceptions of mentoring. Fourteen people participated in this survey.

Student Survey

The students surveyed were from the fall 2007–spring 2008 class of Experimental Projects I and II; the fall 2008 class of Experimental Projects I and II; the fall 2007 and spring 2008 class of Space Systems Engineering; and the fall 2008 class of Flight Vehicle Design, a third capstone course.

Appendix B: Data Collection Instruments

Chapter 1: Writing a Scientific Identity: First Steps

Start-of-Semester Student Interview Questions

1. Who have been your significant writing teachers?

2. What have been your significant writing classes?

3. Describe your writing process for a typical school task—what works best and what do you wish you did differently?

4. What do you see as the relationship between writing/speaking and the work of a biologist/biological engineer?

5. How would you describe the writing and speaking of professionals in your field?

6. What experiences/evidence have created those impressions?

End-of-Semester Student Interview Questions

1. Did you achieve your writing/speaking goals for this class? Why or why not?

2. What aspect of writing/speaking in this class was most difficult for you?

3. What aspect of writing/speaking in this class was easiest for you?

4. If you could have changed anything you did in terms of your writing/speaking in this class, what would it be?

5. If you could have changed anything the instructional staff did in terms of your writing/speaking in this class, what would it be?

6. What kinds of writing/speaking tasks do you feel this class has prepared you for, whether in future coursework or in your career?

7. What kinds of writing/speaking tasks do you wish this class had prepared you for?

Instructional Staff Interview Questions

1. What is your background as a writer, generally, and a scientific writer, specifically?
2. How did you learn to write, particularly in your discipline?
3. How would you describe the strengths of student writers you've worked with at MIT?
4. How would you describe the needs of student writers you've worked with at MIT?
5. What outcomes do you wish for students as a result of the writing/speaking assignments in this class?
6. What do you see as the relationship between laboratory science and writing?

Start-of-Semester Student Survey

Demographics

Your name:

Class standing (check one): ____FR ____SOPH____JUNIOR____SENIOR

Gender: ____M ____F

MIT Major: Course________

Experiences with Scientific/Technical Writing

1. What types of scientific/technical writing have you done (e.g., research articles, lab reports, review articles, etc.)?
2. When you wrote these scientific documents (e.g., research articles, lab reports, technical reports), what did you feel you were learning?
3. What are your strengths as a science or technical writer?
4. What do you struggle with most as a scientific or technical writer?

CI Expectations and Goals

5. Why do you think 7.02 [Introduction to Experimental Biology and Communication] has been designated as a Communications-Intensive class?
6. What sorts of communication skills do biologists need? Which of these do you expect to learn in 7.02?
7. What are your communication goals for 7.02?
8. What kind of writing and speaking do you expect to do after graduation from MIT?

Knowledge of Scientific/Technical Writing

9. Please evaluate the statements below by circling the number corresponding to its importance.

	Extremely Important	Important	Neutral	Slightly Important	Irrelevant
The features of a good *Introduction* include					
Frames research in broader context.	5	4	3	2	1
A single statement of purpose.	5	4	3	2	1
Shows how the work fits in to and extends previous work.	5	4	3	2	1
This sets the direction for your discussion section.	5	4	3	2	1
The features of a good *Methods* include					
Describes all methods used in the research.	5	4	3	2	1
Describes a method for each result or anticipated result.	5	4	3	2	1
Is written before you get the Results.	5	4	3	2	1
The features of good *Results* include					
Support the research question(s).	5	4	3	2	1
Consistent with existing theory.	5	4	3	2	1
Provide unexpected findings.	5	4	3	2	1
Address high profile scientific issue	5	4	3	2	1
Results do not contradict each other, i.e., internally consistent.	5	4	3	2	1
The features of good *Discussions* include					
They summarize findings presented in the results section.	5	4	3	2	1
They cite supporting research literature.	5	4	3	2	1
They explain discrepancies between your findings and previous reports.	5	4	3	2	1
They point out shortcomings of your work and define unsettled points.	5	4	3	2	1
They discuss theoretical and practical implications of your work.	5	4	3	2	1

	Extremely Important	Important	Neutral	Slightly Important	Irrelevant
What are the features of good *Figures*?					
They correspond to key findings.	5	4	3	2	1
They convince reader of your findings (by showing data quality).	5	4	3	2	1
They focus a reader's attention on certain findings (e.g., relationship between values).	5	4	3	2	1
They contain captions that allow readers to make sense of them.	5	4	3	2	1

End-of-Semester Student Survey

Experiences with SciComm and 7.02

1. What do you feel was the most important thing you learned in SciComm?

2. What do you feel was the most *useful* thing you learned in SciComm?

3. When you wrote your SciComm Writing Project, what did you feel you were learning (e.g., format of a research article, science of *Pfu*, experimental methodology)?

4. What do you see as the connection between SciComm and the writing/speaking you will do after MIT?

5. Given your SciComm experience, what do you feel are your strengths as a scientific writer?

6. Given your SciComm experience, what do you feel you struggle with most as a scientific writer?

7. Have your professional goals changed since the beginning of the term? If so, why and what are they now?

Knowledge of Scientific/Technical Writing

8. Please evaluate the statements below by circling the number corresponding to its importance.

	Extremely Important	Important	Neutral	Slightly Important	Irrelevant
The features of a good *Introduction* include					
Frames research in broader context.	5	4	3	2	1
A single statement of purpose.	5	4	3	2	1
Shows how the work fits in to and extends previous work.	5	4	3	2	1
This sets the direction for your discussion section.	5	4	3	2	1
The features of a good *Methods* include					
Describes all methods used in the research.	5	4	3	2	1
Describes a method for each result or anticipated result.	5	4	3	2	1
Is written before you get the Results.	5	4	3	2	1
The features of good *Results* include					
Support the research question(s).	5	4	3	2	1
Consistent with existing theory.	5	4	3	2	1
Provide unexpected findings.	5	4	3	2	1
Address high profile scientific issue	5	4	3	2	1
Results do not contradict each other, i.e., internally consistent.	5	4	3	2	1
The features of good *Discussions* include					
They summarize findings presented in the results section.	5	4	3	2	1
They cite supporting research literature.	5	4	3	2	1
They explain discrepancies between your findings and previous reports.	5	4	3	2	1
They point out shortcomings of your work and define unsettled points.	5	4	3	2	1
They discuss theoretical and practical implications of your work.	5	4	3	2	1

	Extremely Important	Important	Neutral	Slightly Important	Irrelevant
What are the features of good *Figures*?					
They condense large amounts of information.	5	4	3	2	1
They correspond to key findings.	5	4	3	2	1
They convince reader of your findings (by showing data quality).	5	4	3	2	1
They focus a reader's attention on certain findings (e.g., relationship between values).	5	4	3	2	1
They contain captions that allow readers to make sense of them as stand alone entities (i.e., independent of the text).	5	4	3	2	1

Chapter 2: Taking On the Identity of a Professional Researcher

Start of Semester Student Survey

Your name:

Class standing (check one): ____FR ____SOPH____JUNIOR____SENIOR

Gender: ____M ____F

Primary language for writing and speaking:

Please answer the following questions as completely as possible:

1. What types of scientific/technical writing have you done (e.g., research articles, lab reports, review articles, etc.)?
2. Describe a significant writing experience, whether in or out of school. What made it significant for you?
3. What do you struggle with most in your science or engineering writing?
4. What are your strengths as a science or engineering writer?
5. Why do you think 20.109 [Laboratory Fundamentals of Biological Engineering] has been designated as a Communications-Intensive class?
6. What sorts of communication skills do biological engineers need? Which of these do you expect to learn in 20.109?

7. Have you used a Wiki before? What do you see as the function of the Wiki in 20.109?

8. What are your communication goals for 20.109?

9. What kinds of reading and writing do you do outside of school?

10. What kind of writing and speaking do you expect to do after graduation from MIT?

Start-of-Semester Student Interview Questions

1. Who have been your significant writing teachers?

2. What have been your significant writing classes?

3. Describe your writing process for a typical school task—what works best and what do you wish you did differently?

4. What do you see as the relationship between writing/speaking and the work of a biologist/biological engineer?

5. How would you describe the writing and speaking of professionals in your field?

6. What experiences/evidence have created those impressions?

End-of-Semester Student Interview Questions

1. Did you achieve your writing/speaking goals for this class? Why or why not?

2. What aspect of writing/speaking in this class was most difficult for you?

3. What aspect of writing/speaking in this class was easiest for you?

4. If you could have changed anything you did in terms of your writing/speaking in this class, what would it be?

5. If you could have changed anything the instructional staff did in terms of your writing/speaking in this class, what would it be?

6. What kinds of writing/speaking tasks do you feel this class has prepared you for, whether in future coursework or in your career?

7. What kinds of writing/speaking tasks do you wish this class had prepared you for?

Instructional Staff Interview Questions

1. What is your background as a writer, generally, and a scientific writer, specifically?

2. How did you learn to write, particularly in your discipline?

3. How would you describe the strengths of student writers you've worked with at MIT?

4. How would you describe the needs of student writers you've worked with at MIT?

5. What outcomes do you wish for students as a result of the writing/speaking assignments in this class?

6. What do you see as the relationship between laboratory science and writing?

Chapter 3: Carving Out a Research Niche

First Interview Questions

1. Tell me about your project. Why this project? Have you worked in this area before?

2. How are you thinking about that project in terms of writing a grant proposal? How about dividing up that project? What questions to ask/what not to ask? What's "new" or novel about your project? What's too risky for this audience but still interests you?

3. How were you thinking about "pitching" that project—let's walk through section by section

a. Aims—building, crescendo, riskiness?

b. BS [background and significance]—Carving out a niche along the way?

c. What about PD [preliminary data]? What kind of data? How will you use that to build argument?

d. [Research Plan] How will you make reader confident that this project is feasible?

4. As you wrote your grant (to date), to what extent were you thinking about the argument you wanted to make and how best to present it? In other words, to what degree was that "story" central to your writing process? Can you point to specific points in your grant where that was relevant?

5. What advice did you get from the faculty and TAs. What did they say about the argument you were making? Can you point to specific points in your grant where they commented on that issue? How did you address that comment when you revised?

6. What advice did you get from the peer reviewers? What did they say about the argument you were making in your grant? Can you point to specific points in your grant where they commented on that issue? How did you address that comment when you revised?

7. What about your review of peer's grant? What did you learn about making effective arguments by reviewing their work?

8. What do you know about NIH process before this class?

9. What's been new or interesting to learn about?

10. Will you get mentoring from your lab PI or postdoc?

Second Interview Questions

1. Over the semester, how did your thinking about your project in terms of writing a grant proposal change? What do you now think is the most novel thing about your project?

2. What about PD [preliminary data]—What kind did you use? How did you use that to build argument?

3. What about the RP [research plan]—How did you think about writing that section of your grant?

4. Tell me about your experience with the mock study section.

5. How did you approach reviewing a grant? What did you look for? How did you assess it?

6. How would you characterize your responsibilities as a reviewer?

7. What about the review that you received?

8. What's been new or interesting to learn about the NIH process?

9. Will you get other mentoring on grant writing after this class?

10. What was the one thing that you learned in this class?

11. What was the most useful thing that you learned in this class?

12. What was most frustrating? Why?

13. What's to become of your grant? Does it have a life after this class?

TA Interview Questions

1. What do you remember about taking the class? What did you think was the purpose of the class then?

2. What would you say is the purpose of the class now?

3. What's the best thing that you think this year's students got out of the class?

4. What's the most valuable thing that we do in the class?

5. What's one misperception that students have coming into the class?

6. What changes did you see in student writing?

7. What changes did you see in student thinking?

8. At what points did that thinking change?

9. How do you see the class activities contributing to students' understanding of the class goals?

10. What about the study sections?

11. What would you like to do differently?

Faculty Interview Questions

1. What were your ideas behind the CI part of the class? What did you want to accomplish?
2. What's most difficult for students to learn about grant writing?
3. How does this relate to what you teach your own graduate students?
4. How were you thinking about the sequencing of the workshops? Why this order?
5. What did you want students to get out of the study sections?
6. In your experience, what is a faculty member's or project advisor's responsibility in helping a graduate student learn and develop these skills?
7. What's the best thing that you think this year's students got out of the class?
8. What's one misperception?
9. What changes did you see in student writing?
10. What changes did you see in student thinking?
11. At what points did that thinking change?
12. What would you like to do differently?

Frontiers in (Bio)Medical Engineering and Physics: Survey, Spring 2008 (pre)

1. Have you *written* a scientific grant or helped a mentor write a grant? ____ yes ____ no
If yes, what kind of grant? ____
2. Approximately, how many scientific grants have you *read* in the last year?
____ None ____ 1–4 ____ >5
3. Has your current mentor discussed grant writing with you? If yes, in what capacity?
4. Have you started conducting research in your current lab? ____ yes ____ no
5. Structure of a grant: How would you rank the following statements?

	Strongly Agree					Strongly Disagree		
The Introduction of a grant (Overall Goal and Aims) . . .								
may have as many as six measurable aims	7	6	5	4	3	2	1	n/a
includes multiple aims of which the final aim is the most "risky"	7	6	5	4	3	2	1	n/a
should include Aims for which no research has been conducted	7	6	5	4	3	2	1	n/a

	Strongly Agree						Strongly Disagree	
The Background section of a grant . . .								
should show readers that you know how to organize and contextualize research in your field	7	6	5	4	3	2	1	n/a
is primarily descriptive rather than persuasive in function	7	6	5	4	3	2	1	n/a
should include citations from seminal articles and recent research	7	6	5	4	3	2	1	n/a
may include patent references, if appropriate to your research	7	6	5	4	3	2	1	n/a
The Preliminary Data section of a grant . . .								
helps establish your credibility as a researcher	7	6	5	4	3	2	1	n/a
describes *all* data gathered in preliminary experimental research	7	6	5	4	3	2	1	n/a
is organized to match your aims	7	6	5	4	3	2	1	n/a
should include the same amount of preliminary data for each aim	7	6	5	4	3	2	1	n/a
The Research Plan section of a grant . . .								
describes the motivation for your approach(es)	7	6	5	4	3	2	1	n/a
describes expected findings	7	6	5	4	3	2	1	n/a
reads like a lab protocol with specific instructions for completing the experimental methods	7	6	5	4	3	2	1	n/a
is organized by aim	7	6	5	4	3	2	1	n/a

Frontiers in (Bio)Medical Engineering and Physics: Survey Spring 2008 (post)

1. Structure of a grant: How would you rank the following statements?

	Strongly Agree						Strongly Disagree	
The Introduction of a grant (Overall Goal and Aims) . . .								
may have as many as six measurable aims	7	6	5	4	3	2	1	n/a
includes multiple aims of which the final aim is the most "risky"	7	6	5	4	3	2	1	n/a
should include Aims for which no research has been conducted	7	6	5	4	3	2	1	n/a
The Background section of a grant . . .								
should show readers that you know how to organize and contextualize research in your field	7	6	5	4	3	2	1	n/a
is primarily descriptive rather than persuasive in function	7	6	5	4	3	2	1	n/a

	Strongly Agree						Strongly Disagree	
should include citations from seminal articles and recent research	7	6	5	4	3	2	1	n/a
may include patent references, if appropriate to your research	7	6	5	4	3	2	1	n/a
The Preliminary Data section of a grant . . .								
helps establish your credibility as a researcher	7	6	5	4	3	2	1	n/a
describes *all* data gathered in preliminary experimental research	7	6	5	4	3	2	1	n/a
is organized to match your aims	7	6	5	4	3	2	1	n/a
should include the same amount of preliminary data for each aim	7	6	5	4	3	2	1	n/a
The Research Plan section of a grant . . .								
describes the motivation for your approach(es)	7	6	5	4	3	2	1	n/a
describes expected findings	7	6	5	4	3	2	1	n/a
reads like a lab protocol with specific instructions for completing the experimental methods	7	6	5	4	3	2	1	n/a
is organized by aim	7	6	5	4	3	2	1	n/a

Chapter 4: Learning to Argue with Data

Student Interview Questions: First Interviews

1. Tell me about your first project—the microfluidics lab. What did you want to do? Did you find that? What was unexpected? How did you deal with those problems in your draft? Can you point to a specific point in the article where you wrestled with that problem?

2. As you wrote your article/grant, to what extent were you thinking about the argument you wanted to make and how best to present it? In other words, to what degree was that "story" central to your writing process? Can you point to specific points in your article/grant where that was relevant? Did you use a storyboard? Why or why not?

3. The other day we had the Writing Clinic. What advice did you get from the faculty and TAs? What did they say about the argument you were making in your article? What did they say about the figures in your draft? Can you point to specific points in your article/grant where they commented on that issue? How did you address that comment when you revised?

4. What advice did you get from the peer reviewers? What did they say about the argument you were making in your article? What did they say about the figures in your

draft? Can you point to specific points in your article/grant where they commented on that issue? How did you address that comment when you revised?

5. I was really interested in ____________ about your figures. What were you thinking about here as you designed this/these figure(s)?

6. What about your review of the other group? What did you learn about making effective arguments (& figures) by reviewing their article?

7. Tell me about your strategy for revision. In the process of revision, did you find problems with your work? How did you attempt to compensate for anticipated criticism? Can you point to specific points in your article/grant where that was important?

8. Is there anything in your survey answers that you would like to explain or elaborate on more?

Student Interview Questions: Second Interviews

1. Tell me about your second project—the HH [Hodgkin-Huxley] project. What did you want to do? Did you find that?

2. What was unexpected? How did you deal with those problems in your draft? Can you point to a specific point in the article where you wrestled with that problem?

3. As you worked on your presentation, to what extent were you thinking about the argument you wanted to make and how best to present it? In other words, to what degree was that "story" central to your writing process? Can you point to specific points in your presentations where that was relevant?

4. Did you use a storyboard? Why or why not?

5. The other day we had the Writing Clinic. What advice did you get from the faculty and TAs? What did they say about the argument you were making in your presentation?

6. What did they say about the figures in your presentation? Can you point to specific points in your presentations where they commented on that issue?

7. What advice did you get from the peer reviewers?

8. What about your review of the other group?

9. Now that you have done 2 projects, where do you feel that your argument skills could be strengthened? How would that strengthen your technical work?

Faculty Interview Questions

1. In your class, we focus a substantial amount of our "CI" time on the notion of "argument." Can you talk a little bit about what this means to you and what you want students to learn?

2. How do you see the class activities contributing to students' understanding of scientific argument?

3. Have you tried other methods for achieving those ends? What worked/didn't work?

4. What aspects of the lab/communication assignments do you think most contribute to student learning about argument? What do you see as the relationship between the storyboard and argument?

5. Do you think we address the issue of argument suitably in the class? What would you like to do differently?

6. In your experience, what factors make it difficult for a faculty member or project advisor to help an undergraduate student learn and develop these skills?

Quantitative Physiology Survey, Fall 2007 (pre)

Name:

Major: ______________________ (6.1, 6.2, etc.)

1. Have you had other classes that focused on scientific and/or technical writing?

____ No

____ Yes ____ 7.02: Biology Sci Comm

____ 21W. 732: Introduction to Technical Writing

____ Another CIM course at MIT. Which one? ________________

____ Other: ______________________________

If yes, what things did you learn in that class?

____ how to structure a scientific or technical article

____ how to give a scientific or technical talk

____ other forms of scientific or technical communication (e.g., proposals):

____ how to use data to make an argument

____ how to generate scientific and technical figures

____ other: ______________________________

2. Writing and Presenting:

Have you published a scientific or technical article? ____ yes ____ no

Have you given a scientific or technical presentation at a conference? ____ yes ____ no

3. Reading and Listening:

Approximately, how many professional scientific research articles and presentations have you *read or watched* in the last year?

Scientific or technical articles: ____ None ____ 1–4 ____ >5

Scientific or technical presentations (not in class): ____ None ____ 1–4 ____ >

4. Structure of an article: How would you rank the following statements?

	Strongly agree					Strongly disagree		
Writing a technical report or preparing for a presentation can only begin after you are finished with your experiment.	7	6	5	4	3	2	1	n/a
The Introduction of an article should include the hypothesis.	7	6	5	4	3	2	1	n/a
The Methods should include all the methods used in the research.	7	6	5	4	3	2	1	n/a
The Results section should include all the data gathered.	7	6	5	4	3	2	1	n/a
The Results should not include any description of methods.	7	6	5	4	3	2	1	n/a
If the results data do not support the original hypothesis, it means the experiment was not done properly.	7	6	5	4	3	2	1	n/a
Scientific figures should contain as much information as possible.	7	6	5	4	3	2	1	n/a
Scientific figures do not need a caption because they are self-explanatory to readers.	7	6	5	4	3	2	1	n/a
The Discussion should answer your research question.	7	6	5	4	3	2	1	n/a
The Discussion should interpret your results.	7	6	5	4	3	2	1	n/a
The Discussion needs evidence from the results to support interpretations of the data.	7	6	5	4	3	2	1	n/a

5. What do you think is the purpose of the scientific article?

6. Do you think that argument is an important feature in scientific writing? How or how not so?

Quantitative Physiology Survey, Fall 2007 (post)

Name:

1. Have you had a UROP [Undergraduate Research Opportunity]?

2. What do you see as the fundamental differences between presenting data in an oral presentation vs. a written report?

3. At the end of this semester, how have your ideas about the role of argument in scientific writing changed?

4. At the end of this semester, how have your ideas about the visual display of quantitative data in scientific communication changed?

5. Structure of an article: How would you rank the following statements?

	Strongly agree					Strongly disagree		Comments?
Writing a technical report or preparing for a presentation can only begin after you are finished with your experiment.	7	6	5	4	3	2	1	n/a
The Introduction of an article should include the hypothesis.	7	6	5	4	3	2	1	n/a
The Methods should include all methods used in the research.	7	6	5	4	3	2	1	n/a
The Results section should include all the data that you gathered.	7	6	5	4	3	2	1	n/a
The Results should not include any description of methods.	7	6	5	4	3	2	1	n/a
If the results data do not support the original hypothesis, it means the experiment was not done properly.	7	6	5	4	3	2	1	n/a
Scientific figures should contain as much information as possible.	7	6	5	4	3	2	1	n/a
Scientific figures do not need a caption because they are self-explanatory to readers.	7	6	5	4	3	2	1	n/a
The Discussion should answer your research question.	7	6	5	4	3	2	1	n/a
The Discussion should interpret your results.	7	6	5	4	3	2	1	n/a
The Discussion needs evidence from the results to support interpretations of the data.	7	6	5	4	3	2	1	n/a

Chapter 5: Writing and Speaking Collaboratively

Focus Group Surveys and Interview Questions

1. People define team skills in different ways. Think of your own team skills. What were important skills for you on the project?

2. What methods or experiences have been important to you in helping you develop your team skills?

3. What methods or experiences might have helped you learn more about team skills or improve the ones you had?

4. What methods or experiences did NOT help you learn and improve your team skills?

5. Can you summarize the key points of your experience on the project team over the three semesters?

Questions for Interview 1

1. Would you briefly describe the team on which you are working and also tell me something about your project?

2. Who are the people who are part of your team? Are there differences in their roles and responsibilities?

3. Part of your [Experimental Projects I and II] experience is about team skills. In your opinion, what part of your team efforts are going well? What are the strengths you see on your team?

4. In your opinion, what areas or skills could be improved? These can be skills or responsibilities belonging to anyone on your team.

5. In your opinion, how would these changes or improvements strengthen your technical work?

6. How would those changes or improvements strengthen your communication projects (reports, proposals, presentations, briefings)?

Questions for Interview 2

1. What communication deliverables have you produced this term? Not just in 62× but in other courses or activities?

2. How has your process for writing or composing those deliverables changed or stayed the same?

3. You had one collaborative communication project in Experimental Projects I. Can you describe how that differs from your individual communication projects?

Questions for Interview 3

You're in the implementation part of your project: building, testing, taking data.

1. Can you say a little about how the team process has changed or stayed the same?

2. Has your relationship with your partner changed? If so, can you describe this change?

3. Has your sense of who is on your team changed?

4. As you look over a list of skills associated with team skills and collaboration, are there any that seem important to you in this phase of your project?

- interpersonal communication: active listening, giving feedback
- project management: scheduling, meetings, documentation
- leadership: taking initiative for tasks

- followership: allowing others to take responsibility and also being able to implement tasks given to you
- understanding personal psychodynamics and styles of others
- decision-making
- conflict management and resolution

5. You have put together one collaborative deliverable this term: the progress report.
6. Can you talk about that process and how successful you felt it was?

Questions for Interview 4

About the Student

1. You've finished your project. Can you describe how your team skills contributed to this completion phase?
2. Can you describe how your team skills evolved from September '07 to May '08?

About the Student's Teammate

3. Can you describe how your relationship with your partner evolved from September 07 to May 08?
4. What do you know about him/her now that you did not know in the first semester of your project? How did you learn that?
5. Did you observe his/her team skills change or develop in the course of the project?

About the Student's Collaborative Writing Process

6. In this last term, you and your partner composed and presented two oral presentations and a final report. Can you describe your process for doing that work? How did this collaborative process differ from your own individual process? What skills did you find useful? What was not useful?
7. If you were advising 6.2× students on how to best write a collaborative report or give a collaborative talk, what specifically would you tell them?

Other Questions

8. What do you know NOW that you wish you had known about teamwork and collaboration when you began Experimental Projects II?
9. Would you have chosen a different teammate? Would you have insisted on different team processes? If so, what?
10. Could your advisor have given you more support in regard to team processes?
11. Could the faculty have provided more support or instruction in team processes?
12. What else would you like us to know about working as a team to complete a complex project?

Student Survey on Team Skills

Although we ask for your name for record keeping purposes, the data on the survey will be kept confidential. It has no effect on your grade.

In the fall '07, you filled out this survey. We ask you to complete it again. There are no right or wrong answers; just answer based on your current experience and opinions.

These data will be used only to help us learn about how undergraduate engineering students develop their team skills in a project-based engineering course.

1. Team skills can be defined in different ways. In general, what team skills do you think are important in your field of engineering? Please rank the skills listed below, and if you think of other skills not included, please list them in the "other" row.

	Very important 5	4	3	2	Not at all important 1
Interpersonal communication					
▪ active listening					
▪ giving and receiving feedback					
Project management skills					
▪ creating agendas or work breakdown structures					
▪ holding meetings					
▪ documentation					
▪ scheduling and estimating					
Leadership					
▪ taking the initiative for a specific task					
Follower-ship					
▪ implementing team decisions in a timely and thorough way					
Understanding the personal psychodynamics and learning styles of others					
Decision-making					
Conflict management					
▪ Acknowledging disagreements					
▪ Resolving disagreements appropriately					
Other?					

2. From a personal perspective, which team skills have been important to you in your professional development thus far? Again, you may add skills that we have not included here.

	Very important 5	4	3	2	Not at all important 1	Little or no experience
Interpersonal communication						
▪ active listening						
▪ giving and receiving feedback						
Project management skills						
▪ creating agendas or work breakdown structures						
▪ holding meetings						
▪ documentation						
▪ scheduling and estimating						
Leadership						
▪ taking the initiative for a specific task						
Follower-ship						
▪ implementing team decisions in a timely and thorough way						
Understanding the personal psychodynamics and learning styles of others						
Decision-making						
Conflict management						
▪ Acknowledging disagreements						
▪ Resolving disagreements appropriately						
Other?						

3. What methods or experiences have been helpful (or not) to you in developing your team skills? Again, you may add methods that we have not included here.

	Helpful in my development 5	4	3	2	Not helpful in my development 1	Little or no experience
Lectures from faculty (includes readings or handouts)						
Active learning, peer discussion groups, organizational development or "sensitivity" training						
Mentoring from faculty or project/academic advisors or staff						

	Helpful in my development 5	4	3	2	Not helpful in my development 1	Little or no experience
Feedback from peers or faculty or advisors or staff re: team skills						
Experiences outside the classroom (e.g. projects, co-curricular activities, internships, community organizations)						
Modeling: watching other students						
Modeling: watching faculty or advisors or staff						
Gaining insight into the psychodynamics and learning styles of others						
Other:						

4. Focusing more closely on interactions with mentors (faculty or advisors or staff), please indicate the importance of the following activities in helping you develop your team skills.

Even if you have not directly experienced these activities, please rank the importance of the activities according to your opinion. Again, you may add other activities associated with mentoring if they are not included here.

	Very important 5	4	3	2	Not important 1	Little or no experience
Mentor meets face to face and regularly with student.						
Mentor asks challenging questions.						
Mentor gives specific feedback on student's technical work & team skills.						
Mentor does hands-on work with student in lab or shop.						
Mentor offers explanations to student.						
Mentor is available to student by email or during office hours.						
Mentor talks with students about matters other than project or course material (e.g., experiences, activities)						

	Very important 5	4	3	2	Not important 1	Little or no experience
Mentor shares information about professional practices.						
Mentor attends student presentations and/ or team meetings.						
Mentor actively helps student solve technical problems.						
Other:						

5. Could you describe the ways in which your project advisor/mentor was helpful or *not* helpful to you in managing your project and any issues your team experienced? Remember that this information is confidential and has no effect on your grade. For example, you might reflect on some or all of these questions.

- Did your project advisor/mentor make suggestions about how to work as a team?
- Did s/he give suggestions about scheduling and project management?
- Did s/he concentrate solely on technical matters?
- Did you learn about team skills from someone other than your project advisor/ mentor? If so, who?
- Did your project advisor/mentor comment on or review your collaborative presentations before or after you gave them?
- Will s/he help you with your collaboratively written report?

6. If you think that your team skills changed during this two semester course, could you briefly describe specifically which skills changed? What do you think brought that change about?

Faculty Survey on Team Skills

Although we ask for your name for record keeping purposes, the data on the survey will be kept confidential. This data will be used only to help us learn about how undergraduate engineering students develop their team skills in a project-based engineering course.

1. Team skills can be defined in different ways. In general, what team skills do you think are important in your field of engineering or science? Please rank the skills listed below, and if you think of other skills that are not included, please list them in the "other" row.

	Very important 5	4	3	2	Not important 1
Interpersonal communication					
▪ active listening					
▪ giving and receiving feedback					
Project management skills					
▪ creating agendas or work breakdown structures					
▪ holding meetings					
▪ documentation					
▪ scheduling and estimating					
Leadership					
▪ taking the initiative for a specific task					
Follower-ship					
▪ implementing team decisions in a timely and thorough way					
Understanding the personal psychodynamics and learning styles of others					
Decision-making					
Conflict management					
▪ acknowledging disagreements					
▪ resolving disagreements appropriately					
Other?					

2. From a personal perspective, which team skills have been important to you in your professional development thus far? Again, you may add skills that we have not included here.

	Very important 5	4	3	2	Not important 1
Interpersonal communication					
▪ active listening					
▪ giving and receiving feedback					
Project management skills					
▪ creating agendas or work breakdown structures					
▪ holding meetings					
▪ documentation					

	Very important 5	4	3	2	Not important 1
▪ scheduling and estimating					
Leadership					
▪ taking the initiative for a specific task					
Follower-ship					
▪ implementing team decisions in a timely and thorough way					
Understanding the personal psychodynamics and learning styles of others					
Decision-making					
Conflict management					
▪ acknowledging disagreements					
▪ resolving disagreements appropriately					
Other?					

3. This question focuses on what methods or experiences you think are helpful (or not) to your students as they develop their team skills to a professional level. Again, you may add methods that we have not included here.

	Helpful in their development 5	4	3	2	Not helpful in their development 1
Lectures from faculty (includes readings or handouts)					
Active learning, peer discussion groups, organizational development or "sensitivity" training					
Mentoring from faculty or project/academic advisors or staff					
Feedback from peers or faculty or advisors or staff re: team skills					
Experiences outside the classroom (e.g., co-curricular activities, internships, community organizations)					
Modeling: watching other students					
Modeling: watching faculty or advisors or staff					
Gaining insight into the psychodynamics and learning styles of others					
Other:					

4. Focusing more closely on interactions between students and their mentors (faculty or advisors or staff), please indicate the importance of the following activities in helping students develop their *team skills* to a professional level. You may add other activities that we have not included here.

	Very important 5	4	3	2	Not important 1
Mentor meets face to face and regularly with student.					
Mentor asks challenging questions.					
Mentor gives specific feedback on student's technical work and team skills.					
Mentor does hands-on work in lab or shop.					
Mentor offers explanations to student.					
Mentor is available to student by email or during office hours.					
Mentor talks with students about matters other than project or course material. (e.g., hobbies, experiences, activities)					
Mentor shares information about professional practices.					
Mentor attends student presentations and/or team meetings.					
Mentor actively helps student solve technical problems.					
Other:					

Survey on Grading and Assessing Collaborative Communication

1. When I wrote and spoke as part of a TEAM, the faculty assessed our work in the following ways. Please check all that apply.

	Never	Rarely	Sometimes	Often	All the time
Multiple technical grades from professors, mentors and teaching assistants					
Separate communication grade					
Collective grade in which communication and technical grades were merged					
Written feedback					
Verbal feedback					

2. Would you rate the following methods as useful to your learning to write and present? If you did not experience a particular assessment method, please mark N/A.

	Not useful	Rarely useful	Somewhat useful	Useful	Very useful	N/A
Multiple technical grades from professors, mentors and teaching assistants						
Separate communication grade						
Collective grade in which communication and technical grades were merged						
Written feedback						
Verbal feedback						

3. If you received MULTIPLE comments and/or grades on your collaborative documents and reviews, did you find these difficult to understand or at all confusing? If you did not receive multiple comments on your work, please mark N/A.

- not difficult or confusing
- rarely difficult or confusing
- sometimes difficult or confusing
- difficult or confusing
- very difficult or confusing
- N/A

4. If you received MULTIPLE comments and/or grades on your collaborative documents and reviews, did you find this practice useful to your learning? If you did not receive multiple comments on your work, please mark N/A.

- not difficult or confusing
- rarely difficult or confusing
- sometimes difficult or confusing
- difficult or confusing
- very difficult or confusing
- N/A

5. Did you feel that COLLECTIVE grades for collaborative communication were fair to you?

- never

- rarely
- sometimes
- usually
- always
- N/A

6. Specifically how did you use the comments and grades that you and your team received on collaborative communication?

	Never	Rarely	Sometimes	Often	Always
Ignored them					
Reflected on them individually					
Reflected on them with my team					
Hoped others on the team would take action or responsibility					
Talked with a professor, communication instructor or mentor for clarification					
Worked with a communication instructor toward improvement					
Tried to implement feedback in later writing and speaking assignments					

7. What are some of the problems your team encountered in writing and presenting collaboratively?

	Never	Rarely	Sometimes	Usually	Always
Team members did not contribute satisfactory technical content					
Team members did not have strong communication skills					
Team members did not do equal amounts of work					
Team members were not responsible or reliable in producing work					

8. Were some of the following methods useful in helping you and your team learn to write and to present collaboratively? If you did not experience one of these methods, please mark N/A.

	Not useful	Rarely useful	Somewhat useful	Useful	Very useful	N/A
Writing conferences for teams						
Preliminary drafts of design documents or proposals						
Rehearsals ("dry runs") for reviews						
Data analysis workshops						
Seminars or short lectures on collaborative communication strategies						

9. Is there something that we did not ask but that you would like to tell us about the way in which we grade and assess or teach collaborative communication?

References

Accreditation Board for Engineering and Technology. (2008). Criteria for accrediting engineering programs. Retrieved September 25, 2008, from http://www.abet.org/.

Amann, K., & Knorr Cetina, K. (1988). The fixation of (visual) evidence. In M. Lynch & S. Woolgar (Eds.), *Representation in scientific practice* (pp. 85–122). Cambridge, MA: MIT Press.

American Association for the Advancement of Science. (2008). *Project 2061. Benchmarks for science literacy*. Washington, DC: American Association for the Advancement of Science. Retrieved September 25, 2008, from http://www.project2061.org/publications/bsl/online/index.php?chapter=12#D0.

Anderson, P. V. (1985). What survey research tells us. In L. Odell & D. Goswami (Eds.), *Writing in nonacademic settings* (pp. 3–77). New York: Guilford Press.

Anson, C. (2002). *WAC casebook: Scenes for faculty reflection and program development.* New York: Oxford University Press.

Anson, C. (n.d.). Black holes: Writing across the curriculum and the gravitational invisibility of race. In A. Inoue & M. Poe (Eds.), *Race and racism in writing assessment.* Forthcoming.

Anson, C. M., & Forsberg, L. (1990). Moving beyond the academic community: Transitional stages in professional writing. *Written Communication, 7*(2), 200–231.

Artemeva, N. (2005). A time to speak, a time to act: A rhetorical genre analysis of a novice engineer's calculated risk taking. *Journal of Business and Technical Communication, 19*(4), 389–421.

Artemeva, N., Logie, S., & St-Martin, J. (1999). From page to stage: How theories of genre and situated learning help introduce engineering students to discipline-specific communication. *Technical Communications Quarterly, 3*, 301–316.

Baigrie, B. (1996). *Picturing knowledge: Historical and philosophical problems concerning the use of art in science.* Toronto: University of Toronto Press.

Bangert-Downs, R., Hurley, M., & Wilkinson, B. (2004). The effects of school-based writing-to-learn interventions on academic achievement: A meta-analysis. *Review of Educational Research, 74*(1), 29–58.

Bartholomae, D. (1986). Inventing the university. *Journal of Basic Writing, 5*, 4–23.

Baxter Magolda, M. B. (1999). *Creating context for learning and self-authorship: Constructive-developmental pedagogy*. Nashville, TN: Vanderbilt University Press.

Bazerman, C. (1988). *Shaping written knowledge: The genre and activity of the experimental article in science*. Madison: University of Wisconsin Press.

Bazerman, C. (2004). Speech acts, genres, and activity systems: How texts organize activity and people. In C. Bazerman & P. Prior (Eds.), *What writing does and how it does it: An introduction to analyzing texts and textual practices* (pp. 309–340). Mahwah, NJ: Erlbaum.

Bazerman, C., Little, J., Bethel, L., Chavkin, T., Fouquette, D., & Garufis, J. (2005). *Reference guide to writing across the curriculum*. West Lafayette, IN: Parlor Press.

Bazerman, C., & Paradis, J. (1991). *Textual dynamics of the professions*. Madison: University of Wisconsin Press.

Bean, J. (2008, May). *Making data speak: Border crossings in rhetoric, chemistry, literature, and mathematics*. Presentation at Ninth Biennial International Writing Across the Curriculum Conference, Austin, TX.

Beaufort, A. (2007). *College writing and beyond: A new framework for university writing instruction*. Logan: Utah State University Press.

Berkenkotter, C., & Huckin, T. (1995). *Genre knowledge in disciplinary communication*. Mahwah, NJ: Erlbaum.

Berkenkotter, C., Huckin, T., & Ackerman, J. (1991). Social contexts and socially constructed texts. The initiation of a graduate student into a writing research community. In C. Bazerman & J. Paradis (Eds.), *Textual dynamics of the professions: Historical and contemporary studies of writing in academic and other professional communities* (pp. 191–215). Madison: University of Wisconsin Press.

Bizzell, P. (1992). *Academic discourse and critical consciousness*. Pittsburgh, PA: University of Pittsburgh Press.

Blakeslee, A. (1997). Activity, context, interaction, and authority: Learning to write scientific papers in situ. *Journal of Business and Technical Communication, 11*(2), 125–169.

Blakeslee, A. (2001). *Interacting with audiences: Social influences on the production of scientific writing*. Mahwah, NJ: Lawrence Erlbaum Associates.

Bloom, B., & Krathwohl, D. R. (1956). *Taxonomy of educational objectives: The classification of educational goals, by a committee of college and university examiners. Handbook 1: Cognitive domain*. New York: Longman.

Bransford, J. (2001). *Thoughts on adaptive expertise*. Retrieved January 12, 2009, from http://www.vanth.org/docs/AdaptiveExpertise.pdf.

Bransford, J., Brown, A., & Cocking, R. (2000). *How people learn: Brain, mind, experience, and school*. Washington, DC: National Academy Press.

Brown, B. A., Reveles, J. M., & Kelly, G. J. (2005). Scientific literacy and discursive identity: A theoretical framework for understanding science learning. *Science Education, 89*, 779–802.

Bruffee, K. A. (1984). Collaborative learning and the "conversation of mankind." *College English, 46*(7), 635–652.

Burrell, B. D., & Colton, C. K. (1999, June). *How to initiate dialogue in student research teams*. Presentation at American Society for Engineering Education, Charlotte, NC.

Campbell, J. P. (1893). *Biological teaching in the colleges of the United States*. Washington, DC: Government Printing Office.

Carroll, L. A. (2002). *Rehearsing new roles: How college students develop as writers*. Carbondale: Southern Illinois University Press.

Carter, M., Ferzli, M., & Wiebe, E. N. (2007). Writing to learn by learning to write in the disciplines. *Journal of Business and Technical Communication, 21*(1), 278–302.

CDIO syllabus report. (2001). Cambridge, MA: Department of Aeronautics and Astronautics, MIT.

Chan, L. Y., Kosuri, S., & Endy, D. (2005). Refactoring bacteriophage T7. *Molecular Systems Biology*, 10.1038, E1–E10.

Chisholm, R. (1990). Coping with the problems of collaborative writing. *Writing Across the Curriculum, 2*, 90–108.

Cochran-Smith, M., & Lytle, S. L. (1993). *Inside/Outside: Teacher research and knowledge*. New York: Teachers College Press.

Collins, A., Brown, J. S., & Newman, S. E. (1989). Cognitive apprenticeship: Teaching the crafts of reading, writing, and mathematics. In L. B. Resnick (Ed.), *Knowing, learning, and instruction: Essays in honor of Robert Glaser* (pp. 453–494). Mahwah, NJ: Erlbaum.

Connor, U., & Upton, T. (1996). The genre of grant proposals: A corpus linguistics analysis. In U. Connor & T. Upton (Eds.), *Discourse in the professions: Perspectives from corpus linguistics* (pp. 235–256). Amsterdam: John Benjamins Publishing.

Craig, J. L., & Coleman, C. (2004, June). *Using teamwork and communication skills to monitor and strengthen the effectiveness of undergraduate aerospace engineering design projects*. Presentation at American Society of Engineering Education, Salt Lake City, UT.

Dannels, D. P. (2000). Learning to be professional: Technical classroom discourse, practice, and professional identity construction. *Journal of Business and Technical Communication, 14*(1), 5–32.

Department of Aeronautics and Astronautics. (n.d.). *MIT Department of Aeronautics and Astronautics*. (online), Retrieved August 11, 2008, from http://web.mit.edu/aeroastro/index.html.

Devitt, A. (2004). *Writing genres*. Carbondale: Southern Illinois Press.

Ding, H. (2008). The use of cognitive and social apprenticeship to teach a disciplinary genre. *Written Communication, 25*(1), 3–52.

Driskill, L. (2000). *Linking industry best practice and EC3(g) assessment in engineering communication.* Paper presented at the American Society for Engineering Education conference, St. Louis, MO.

Driver, R., Newton, P., & Osborne, J. (1997). Establishing the norms off scientific argumentation in classrooms. *Science Education, 84*(3), 287–312.

Ede, L., & Lunsford, A. (1983). Why write . . . together? *Rhetoric Review, 1*(2), 150–157.

Ede, L., & Lunsford, A. (2001). Collaboration and concepts of authority. *PMLA, 116*(2), 354–369.

Elbow, P., & Belanoff, P. (2000). *Sharing and responding* (3rd ed.). New York: McGraw-Hill.

Ellis, R. A. (2004). University student approaches to learning science through writing. *International Journal of Science Education, 26*(15), 1835–1853.

Ellis, R. A., Taylor, C. E., & Drury, H. (2006). University student conceptions of learning science through writing. *Australian Journal of Education, 50*(1), 6–28.

Experimental projects I and II course syllabus. (2008). Cambridge, MA: Department of Aeronautics and Astronautics, MIT.

Facione, Peter A. (n.d.) *Critical thinking: What it is and why it counts.* Retrieved November 25, 2003, from www.insightassessment.com/pdf_files/what&why2006.pdf.

Faigley, L., & Miller, T. P. (1982). What we learn from writing on the job. *College English, 44*(6), 557–569.

Finn, R. (1995). NIH study section members acknowledge major flaws in the reviewing system. *Scientist, 9*(16), 7.

Freedman, A., & Adam, C. (1996). Learning to write professionally: "Situated learning" and the transition from university to professional discourse. *Journal of Business and Technical Communication, 10*(4), 395–427.

Freedman, A., Adam, C., & Smart, G. (1994). Wearing suits to class: Simulating genres and simulations as genre. *Written Communication, 11*(2), 193–226.

Gee, J. P. (2000). The new literacy studies: From "socially situated" to the work of the social. In D. Barton, M. Hamilton, & R. Ivanic (Eds.), *Situated literacies: Reading and writing in context* (pp. 180–209). London: Routledge.

Geller, A. E. (2005). "What's cool here?" Collaboratively learning genre in biology. In A. Herrington & C. Moran (Eds.), *Genre across the curriculum* (pp. 83–105). Logan: Utah State University Press.

Gladfelter, E. H. (2002). *Agassiz's legacy: Scientists' reflections on the value of field experience.* New York: Oxford University Press.

Greitzer, E. M. (2007). *Some aerodynamic problems of aircraft engines—fifty years after.* Presentation at the ASME Turbo Expo 2007: Power for Land, Sea, and Air, Montreal, Canada.

Grier, N. M. (1935). On the approval for accrediment of college science laboratories. *Science Education, 19*(1), 19–23.

Gross, A. (1990). *The rhetoric of science.* Cambridge, MA: Harvard University Press.

Gross, A. (2006). *Starring the text: The place of rhetoric in science studies.* Carbondale: Southern Illinois University Press.

Gross, A., Harmon, J., & Reidy, M. (2002). *Communicating science: The scientific article from the seventeenth century to the present.* New York: Oxford University Press.

Gurak, L., & Lannon, J. (2001). *A concise guide to technical communication.* New York: Longman.

Haas, C. (1994). Learning to read biology: One student's rhetorical development in college. *Written Communication, 11,* 43–84.

Halpern, D. F. (1998). Teaching critical thinking for transfer across domains: Dispositions, skills, structure training, and metacognitive monitoring, *American Psychologist, 53*(4), 449–455.

Haswell, R. H. (1991). *Gaining grounding ground in college writing: Tales of development and interpretation.* Dallas: Southern Methodist University Press.

Herrington, A. (1985). Writing in academic settings: A study of the contexts for writing in two college chemical engineering courses. *Research in the Teaching of English, 19*(4), 331–361.

Hodgkin-Huxley project guidelines. (2007, fall). *6.021J Quantitative physiology: Cells and tissues.* Cambridge, MA: MIT.

Hodson, D. (1998). Is this really what scientists do? Seeking a more authentic science and beyond the school laboratory. In J. Wellington (Ed.), *Practical work in school science: Which way now?* (pp. 93–108). London: Routledge.

How to educate young scientists. (2006, July 3). *New York Times,* p. A14.

HST 500 Course syllabus (2008, spring). Health Sciences Technology. Cambridge, MA: MIT.

Hyland, K. (2004). *Disciplinary discourses: Social interactions in academic writing.* Ann Arbor: University of Michigan Press.

Inoue, A. B. (2005). Community-based assessment pedagogy. *Assessing Writing, 9,* 208–238.

Inouye, S., & Fiellin, D. (2005). An evidence-based guide to writing grant proposals for clinical research. *Annals of Internal Medicine, 14*(4), 27–282.

An introduction to writing across the curriculum. (1997). Retrieved January 12, 2009, from http://wac.colostate.edu/intro/.

Katzenbach, J. R., & Smith, D. K. (1993). The discipline of teams. *Harvard Business Review, 71*(2), 111–120.

Kaufman, D. B., Felder, R. M., & Fuller, H. (2000). Accounting for individual effort in cooperative learning teams. *Journal of Engineering Education, 89*(2), 133–140.

Keys, C. (1999). Revitalizing instruction in scientific genres: Connecting knowledge production with writing to learn in science. *Science Education, 83,* 115–130.

King, P., & Kitchener, K. (1994). *Developing reflective judgment: Understanding and promoting intellectual growth and critical thinking in adolescents and adults.* San Francisco: Jossey-Bass.

Klein, P. D. (1999). Reopening inquiry into cognitive processes in writing-to-learn. *Educational Psychology Review, 11*(3), 203–270.

Knorr Cetina, K. (1981). *The manufacture of knowledge: An essay on the constructivist and contextual nature of science.* Oxford: Pergamon Press.

Knorr Cetina, K. (1999). *Epistemic cultures: How the sciences make knowledge.* Cambridge, MA: Harvard University Press.

Kuldell, N. (2007). *Lecture 1 notes, 20.109 Laboratory Fundamentals of Biological Engineering.* Cambridge, MA: MIT.

Langer, J. A., & Applebee, A. N. (1987). *How writing shapes thinking: A study of teaching and learning.* Urbana, IL: NCTE.

Latour, B. (1987). *Science in action: How to follow scientists and engineers through society.* Cambridge, MA: Harvard University Press.

Latour, B., & Woolgar, S. (1979). *Laboratory life: The construction of scientific facts.* Princeton, NJ: Princeton University Press.

Lave, J. (1996). Teaching, as learning, in practice. *Mind, Culture, and Activity, 3*(3), 149–164.

Lave, J., & Wenger, E. (1991). *Situated learning: Legitimate peripheral participation.* Cambridge: Cambridge University Press.

Lerner, N. (2007). Laboratory lessons for writing and science. *Written Communication, 24*(3), 191–222.

Lewis, P., Aldridge, D., & Swamidass, P. M. (1998). Assessing teaming skills acquisition on undergraduate project teams. *Journal of Engineering Education, 87*(2), 149–155.

Leydens, J. A. (2008). Novice and insider perspectives on academic workplace writing: Toward a continuum of rhetorical awareness. *IEEE Transactions on Professional Communication, 51*(3), 242–263.

Leydens, J., & Olds, B. (2007). Publishing in scientific and engineering contexts: A course for graduate students. *IEEE Transactions on Professional Communication, 50*(1), 45–56.

Locke, D. (1992). *Science as writing.* New Haven, CT: Yale University Press.

Loughry, M. L., Ohland, M. H., & Moore, D. D. (2007). Development of a theory-based assessment of team member effectiveness. *Educational and Psychological Measurement, 67,* 505–524.

Louth, R. (1989). Introducing collaborative writing. In R. Louth & A. M. Scott (Eds.), *Collaborative technical writing: Theory and practice* (pp. 1–7). Hammond, IN: Association of Teachers of Technical Writing.

Lovgren, R. H., & Racer, M. J. (2000). Group dynamics in projects: Don't forget the social aspects. *Journal of Professional Issues in Engineering Education and Practice, 126*(4), 156–165.

Lowry, P. B., Nunamaker, J. F., Curtis, A., & Lowry, M. R. (2005). The impact of process structure on novice, virtual collaboration writing teams. *IEEE Transactions on Professional Communication, 48*(4), 341–364.

Luzon, M. J. (2005). Genre analysis in technical communication. *IEEE Transactions on Professional Communication, 48*(3), 285–295.

Lynch, M. (2006a). Discipline and the material form of images: An analysis of scientific visibility. In L. Pauwels (Ed.), *Visual cultures of science: Rethinking representational practices in knowledge building and science communication* (pp. 195–21). Hanover, NH: Dartmouth College Press.

Lynch, M. (2006b). The production of scientific images: Vision and re-vision in the history, philosophy, and sociology of science. In L. Pauwels (Ed.), *Visual cultures of science: Rethinking representational practices in knowledge building and science communication* (pp. 26–40). Hanover, NH: Dartmouth College Press.

Lynch, M., & Woolgar, S. (1990). *Representation in scientific practice.* Cambridge, MA: MIT Press.

MacIntosh-Murray, A. (2007). Poster presentations as a genre in knowledge construction: A case study of forms, norms, and values. *Science Communication, 28*, 347–376.

Materials Research Laboratory, University of California, Santa Barbara. (n.d.). *Education Outreach Program.* Retrieved September 25, 2008, from http://btc.mrl.ucsb.edu/mrl/outreach/educational/index.html.

Mathison, M. (2000). "I don't have to argue my design—the visual speaks for itself: A case study of mediated activity in an introductory mechanical engineering course." In S. Mitchell & R. Andrews (Eds.), *Learning to argue in higher education* (pp. 74–84). Portsmouth, NH: Boynton/Cook Heinemann.

McCarthy, L. P. (1987). Stranger in strange lands: A college student writing across the curriculum. *Research in the Teaching of English, 21*(3), 233–265.

McGinn, M., & Roth, W. (1999). Preparing students for competent scientific practice: Implications of recent research in science and technology studies. *Educational Researcher, 28*(3), 14–24.

McGourty, J., & De Meuse, K. P. (2001). *The team developer.* Hoboken, NJ: Wiley.

McLeod, S. E., Soven, M., & Thaiss, C. (2001). *WAC for the new millennium: Strategies for continuing writing-across-the-curriculum programs.* Urbana, IL: NCTE Press.

Michaelson, M. (1990). How an author can avoid the pitfalls of practical ethics. *IEEE Transactions in Professional Communication, 33*(2), 58–61.

Microfluidics Project Overview. (2007, fall). 6.021J Quantitative Physiology: Cells and Tissues. Cambridge, MA: MIT.

Miller, C. R. (1984). Genre as social action. *Quarterly Journal of Speech, 70*, 151–176.

Miller, C. R. (1979). A humanistic rationale for technical writing. *College English, 40*(6), 610–617.

MIT. (2008). Teaching and Learning Laboratory and the faculty subcommittee on the communications requirement. In *Implementation of the undergraduate communication requirement: A report on the assessment* (2005–2007). Cambridge, MA: MIT.

MIT Biological Engineering. (2003). *Educational programs: BE SB degree requirements (2009+).* Retrieved September 5, 2008, from http://web.mit.edu/be/education/ugrad-reqs_2009.htm.

MIT bulletin: Course catalogue. (2008). Cambridge, MA: MIT.

MIT facts. (2007). Retrieved January 12, 2009, from http://web.mit.edu/facts/ enrollment.html.

Monroe, J. (2002). *Writing and revising the disciplines.* Ithaca, NY: Cornell University Press.

Myers, G. (1990). *Writing biology: Texts in the social construction of scientific knowledge.* Madison: University of Wisconsin Press.

National Institutes of Health. (2004). *NIH announces updated criteria for evaluating research grant applications* (online). Retrieved August 1, 2008, from http://grants.nih.gov/granTs/guide/notice-files/NOT-OD-05-002.HTML.

National Institutes of Health. (2008a). *Estimates of funding for various diseases, conditions, research areas.* Retrieved August 1, 2008, from http://www.nih.gov/news/fundingresearchareas.htm.

National Institutes of Health. (2008b). *Research grant program.* Retrieved December 23, 2008, from http://grants.nih.gov/grants/funding/r01.htm.

Oakley, B., Felder, R. M., Brent, R., & Elhajj, I. (2004). Turning student groups into effective teams. *Journal of Student Centered Learning, 2*(1), 9–34.

Ohland, M. W., Layton, R. A., Loughry, M. L., & Yuhasz, A. G. (2005). Effects of behavioral anchors on peer evaluation reliability. *Journal of Engineering Education, 94*(3), 319–326.

Oschsner, R., & Fowler, J. (2004). Playing devil's advocate: Evaluating the literature of the WAC/WID movement. *Review of Educational Research, 74*(2), 117–140.

Office of the Communication Requirement. (2008). *About the requirement: MIT undergraduate communication requirement.* Retrieved February 1, 2008, from http://web.mit.edu/commreq/background.html.

Pappas, E. C., & Hendricks, R. W. (2000). Holistic grading in science and engineering. *Journal of Engineering Education, 90*(4), 403–408.

Paradis, J., & Dobrin, D. (1985). Writing at Exxon ITD: Notes on the writing environment of an R & D organization. In Odell, L. & Goswami D. (Eds.), *Writing in nonacademic settings* (pp. 281–307). New York: Guilford Press.

Paretti, M. C., & McNair, L. D. (2008). Introduction to special issue on communication in engineering curricula: Mapping the landscape. *IEEE Transactions on Professional Communication, 51*(3), 238–241.

Paris, S. G., Lipson, M. Y., & Wixson, K. K. (1983). Becoming a strategic reader. *Contemporary Educational Psychology, 8*, 293–316.

Paris, S. G., & Winograd, P. (1990). How metacognition can promote academic learning and instruction. In L. I. Beau Fly Jones (Ed.), *Dimensions of thinking and cognitive instruction* (pp. 15–51). Mahwah, NJ: Erlbaum.

Patton, M., & Nagelhout, E. (2004). Literacy and learning in context: Biology students in the classroom and the lab. In B. Huot, B. Stroble, & C. Bazerman (Eds.), *Multiple literacies for the 21st century* (pp. 151–171). Cresskill, NJ: Hampton.

Pauwels, L. (2006). *Visual cultures of science: Rethinking representational practices in knowledge building and science communication.* Hanover, NH: Dartmouth College Press.

Penrose, A., & Katz, S. (1997). *Writing in the Sciences: Exploring Conventions of Scientific Discourse.* New York, NY: Pearson Longman.

Perelman, L., Paradis, J., & Barrett, E. (1997). *Mayfield handbook of scientific and technical writing.* New York: McGraw-Hill.

Perry, W. (1970). *Forms of intellectual and ethical development in the college years: A scheme.* New York: Holt.

Porter, J. E. (1992). *Audience and rhetoric.* Englewood Cliffs, NJ: Prentice Hall.

Porush, D. (1995). *A short guide to writing about biology.* New York: Longman.

Prelli, L. (1990). *A rhetoric of science: Inventing scientific discourse.* Columbia: University of South Carolina Press.

Prince, M. (2004). Does active learning work? A review of the research. *Journal of Engineering Education, 93*(3), 223–231.

Prior, P. (2006). A sociocultural theory of writing. In C. MacArthur, S. Graham, & J. Fitzgerald (Eds.), *Handbook of writing research* (pp. 275–292). New York: Guilford Press.

Pritchard, R., & Honeycutt, R. (2006). The process approach to writing instruction. In C. MacArthur, S. Graham, & J. Fitzgerald (Eds.), *Handbook of writing research* (pp. 275–292). New York: Guilford Press.

Resnik, D., Shamoo, A., & Krimsky, S. (2006). Fraudulent human embryonic stem cell research in South Korea: Lessons learned. *Accountability in Research, 13*(1), 101–109.

Riel, M. (1998). *Education in the 21st century: Just-in-time learning or learning communities.* Presentation at the Fourth Annual Conference for the Emirates Center for Strategic Studies and Research, Abu Dhabi.

Rogoff, B. (1995). Observing sociocultural activity on three planes: Participatory appropriation, guided participation, and apprenticeship. In J. V. Wertsch, P. del Rio, & A. Alvarez (Eds.), *Sociocultural studies of the mind* (pp. 139–164). Cambridge: Cambridge University Press, 1995.

Roth, W., Pozzer-Ardenghi, L., & Han, J. Y. (2005). *Critical graphicacy: Understanding visual representation practices in school science*. Dordrecht, Netherlands: Springer.

Russell, C., & Weaver, G. (2008). Student perceptions of the purpose and functions of the laboratory in science: A grounded theory study. *International Journal for the Scholarship of Teaching and Learning, 2*(2), n.p. Retrieved September 26, 2008, from http://academics.georgiasouthern.edu/ijsotl/v2n2/articles/_Russell/index.htms.

Russell, D. R. (1997). Writing and genre in higher education and workplaces: A review of studies that use cultural-historical activity theory. *Mind, Culture, and Activity, 4*(4), 224–237.

Russell, D. R. (2002). *Writing in the academic disciplines: A curricular history*, Carbondale: Southern Illinois University Press. (Original work published 1991)

SciComm syllabus. (2008, spring). Cambridge, MA: Department of Biology, MIT.

Schulz, K. H., & Ludlow, D. K. (1996, July). Incorporating group writing instruction in engineering courses. *Journal of Engineering Education*, 227–232.

Segall, M., & Smart, R. (2005). *Direct from the disciplines: Writing across the curriculum*. Portsmouth, NH: Boynton/Cook.

7.02 manual (2007, fall). Cambridge, MA: Department of Biology, MIT.

Shuman, L., Besterfield-Sacre, M., & McGourty, J. (2005). The ABET "professional skills"—Can they be taught? Can they be assessed? *Journal of Engineering Education, 94*(1), 41–55.

6.021J course syllabus (Fall 2007). Cambridge, MA: Department of Electrical Engineering and Computer Science, MIT.

Smith, K. A. (1995) *Cooperative learning: Effective teamwork for engineering classrooms*. Frontiers in Education Conference Proceedings, Atlanta, GA.

Smith, K. A., & Imbrie, P. K. (2007). *Teamwork and project management* (3rd ed.). New York: McGraw-Hill.

Smith, K. A., Sheppard, S. D., Johnson, D. W., & Johnson, R. T. (2005). Pedagogies of engagement: Classroom-based practices. *Journal of Engineering Education, 94*(1), 87–101.

Smith, M., Cheville, J., & Hilloks, G. (2006). I guess I'd better watch my English. In C. MacArthur, S. Graham, & J. Fitzgerald (Eds.), *Handbook of writing research* (pp. 263–274). New York: Guilford Press.

Sommers, N. (2008). The call of research: A longitudinal view of writing development. *College Composition and Communication, 60*(1), 152–164.

Space Systems Engineering course syllabus. (Spring 2006). Massachusetts Institute of Technology, Department of Aeronautics and Astronautics, Cambridge, MA: MIT.

Speck, B. W., Johnson, T. R., & Heaton, L. B. (1999). *Collaborative writing: An annotated bibliography*. Westport, CT: Greenwood Press.

Spinuzzi, C. (2003). *Tracing genres through organizations: A sociocultural approach to information design*. Cambridge, MA: MIT Press.

Springer, L., Stanne, M. E., & Donovan, S. S. (1999). Effects of small-group learning on undergraduates in science, mathematics, engineering and technology: A meta-analysis. *Review of Educational Research, 69*(1), 21–51.

Sternglass, M. S. (1997). *Time to known them: A longitudinal study of writing and learning at the college level*. Mahwah, NJ: Erlbaum.

Stratton, C. R. (1989). Collaborative writing in the workplace. *IEEE Transactions in Professional Communication, 32*(3), 178–182.

Swales, J. M. (1990). *Genre analysis: English in academic and research settings*. Cambridge: Cambridge University Press.

Swales, J. (2004). *Research genres: Explorations and applications*. Cambridge: Cambridge University Press.

Thaiss, C., & Zawacki, T. M. (2006). *Engaged writers, dynamic disciplines: Research on the academic writing life*. Portsmouth: Boynton/Cook.

Trafton, A. (2008). Wave of the future: MIT to graduate first class of biological engineering majors. *MIT News*. Retrieved September 5, 2008, from http://web.mit.edu/newsoffice/2008/bioengineering-0602.html.

Tufte, E. (2001). *The visual display of quantitative information*. Cheshire, CT: Graphics Press. (Original work published 1984.)

Tynjala, P., Mason, L., & Loonka, K. (2001). *Writing as a learning tool: Integrating theory and practice*. Dordrecht: Kluwer.

Vygotsky, L. S. (1978). *Mind in society: The development of higher psychological processes*. Cambridge, MA: Harvard University Press.

Walker, K. (1999). Using genre theory to teach students engineering lab report writing: A collaborative approach. *IEEE Transactions on Professional Communication, 42*(1), 12–19.

Walvoord, B. E., & McCarthy, L. P. (1990). *Thinking and writing in college: A naturalistic study of students in four disciplines*. Urbana, IL: NCTE.

Wilkerson, L., & Abelman, W. (1993). Producing physician-scientists: A survey of graduates from the Harvard-MIT program in health sciences technology. *Academic Medicine, 68*(3), 214–218.

Winsor, D. A. (1996). *Writing like an engineer: A rhetorical education*. Mahwah, NJ: Erlbaum.

Winsor, D. A. (2003). *Writing power: Communication in an engineering center*. Albany: State University of New York Albany Press.

Wolfe, J., & Alexander, K. (2005). The computer-expert in the mixed-gendered technical writing group. *Journal of Business and Technical Communication, 19*(2), 135–170.

Wood, D., Bruner, J. S., & Ross, G. (1976). The role of tutoring in problem solving. *Journal of Child Psychology and Psychiatry*, *17*, 89–100.

Yancey, K., & Huot, B. (1997). *Assessing writing across the curriculum*. Greenwich, CT: Ablex.

Zerbe, M. J. (2007). *Composition and the rhetoric of Science: Engaging the dominant discourse*. Carbondale: Southern Illinois University Press.

Zerhouni, E. (2006). NIH in the post-doubling era: Realities and strategies, *Science*, *17*, 1088–1090.

Index